ASHEVILLE-BUNCOMBE TECHNICAL INSTITUTE

NORTH CAROLINA
STATE BOARD OF EDUCATION
DEPT. OF COMMUNITY COLLEGES
LIBRARIES

DISCARDED

JUN 2 6 2025

D1526050

manufacture
and
analysis of
CARBONATED BEVERAGES

MORRIS B. JACOBS, Ph.D.
*Consulting Chemist
and Microbiologist
and Professional
Engineer*

1959
CHEMICAL PUBLISHING CO., INC.

212 Fifth Avenue　　　　　　　　　　　　New York, N. Y.

© 1959
CHEMICAL PUBLISHING CO., INC.
212 Fifth Avenue, New York, N. Y.

Printed in the United States of America

PREFACE

In this book I have endeavored to present a comprehensive treatment of the manufacture and analysis of carbonated nonalcoholic beverages or carbonated soft drinks as they are commonly called. Each category of the raw materials used in these beverages is considered, namely, sugars and sirups, artificial sweetening agents, acids, water, flavors and flavoring, including specialty and fruit flavors and also flavor emulsions both of the clear and cloudy type, colors and coloring, and carbon dioxide. The actual manufacturing steps are described in detail in the chapter on bottling and canning and, in this connection, bottle washing, caustic solution preparation, plant layout, plant housekeeping, and sanitation are discussed in detail. The composition of the finished beverages by categories is considered. The various types of spoilage that may occur and the means for the prevention of such spoilage are treated in another chapter. Finally the methods of analysis both for control during manufacture and for the determination of composition are detailed.

I acknowledge with thanks the cooperation given to me by the American Bottlers of Carbonated Beverages and by several firms. These acknowledgements are given specifically in the text. It may be noted that reference is made to American Bottlers of Carbonated Beverages, the United States Pharmacopeia, and the National Formulary standards and specifications for a number of raw materials. These agencies speak for themselves and their standards are quoted merely as guides for desirable practice.

Some beverage, flavor, and color formulations are listed in this book. Some of the compositions mentioned are illustrative of commercial practice while others are of an experimental nature. These formulas have been included to serve as suggestions to the manufacturer; they are not given as a formulary. The application of knowledge of the art and skill may result in the improvement of these formulations.

The mention of a particular substance in this book does not in any manner imply that I approve of the use of such a substance.

Bayside, New York   MORRIS B. JACOBS

TABLE OF CONTENTS

Chapter
1. INTRODUCTION   13
    1. Definitions
    2. Historical Development
        a. Mineral-Water Period
        b. Soda-Water Period
    3. Production and Sales

2. SUGARS AND SIRUPS   26
    1. Sugars
        a. Sucrose
        b. Liquid Sugar
        c. Invert Sugar and Sirups
        d. Dextrose Sugars
    2. Beverage-Sirup Manufacture
        a. Simple Sirup
        b. Cold Process Sirup
        c. Acidified Cold Process Sirup
        d. Hot Process Sirup
        e. Acidified Hot Process Sirup
        f. High Density Sirup
    3. Sugar Concentration and Dilution
    4. Beverage-Sugar Concentration

3. ARTIFICIAL AND HIGH-POWER SWEETENING AGENTS   49
    A. Artificial Sweeteners
        1. Saccharin
            a. U.S.P. Requirements for Saccharin

Chapter
3. ARTIFICIAL AND HIGH-POWER SWEETENING
   AGENTS—*Continued* 49
   A. Artificial Sweeteners—*Continued*
      1. Saccharin—*Continued*
         b. U.S.P. Requirements for Soluble Saccharin
         c. Formulation
      2. Cyclamate
         a. Sodium Cyclamate
         b. Calcium Cyclamate
         c. Other Sulfamates
         d. Toxicity
         e. Formulation
      3. Dulcin
      4. Alkoxyaminonitrobenzenes
   B. Natural High-Power Sweetening Agents
      1. Perilla Aldehyde Antialdoxime
      2. Stevioside

4. ACIDS AND ACIDULATION 68
   1. Acidity
      a. pH
      b. Normality
   2. Citric Acid
      a. U.S.P. Requirements
   3. Tartaric Acid
      a. N.F. Requirements
   4. Malic Acid
   5. Lactic Acid
      a. U.S.P. Requirements
   6. Acetic Acid
   7. Phosphoric Acid
      a. U.S.P. Requirements
   8. Acidulation

5. WATER AND WATER TREATMENT 89
   1. Criteria
   2. Classification
   3. Water Requirements
      a. Clear Water

Chapter

5. WATER AND WATER TREATMENT—*Continued*   89
   3. Water Requirements—*Continued*
      b. Colorless Water
      c. Tasteless Water
      d. Odorless Water
      e. Iron
      f. Organic Matter
      g. Low-Alkalinity Water
      h. Sterile Water
   4. Water Treatment
      a. Permutit Precipitator
      b. Electrolytic Coagulation
      c. Cation Exchange

6. FLAVORS AND FLAVORING   110
   1. Flavor Sensation
      a. Sugar
      b. Artificial Sweeteners
      c. Acid
      d. Flavors
   2. Classification of Flavoring Materials
      a. Naturally Occurring Flavoring Materials
      b. Inorganic Natural Flavoring Materials
      c. Natural Flavors
      d. Isolates
      e. Semisynthetics
      f. Synthetics
   3. Criteria for Beverage Flavors
      a. Solubility
      b. Flavor Fidelity
      c. Acid Resistance
      d. Noncontamination
      e. Other Criteria
   4. Natural Flavorings
      a. Essential Oils
      b. Oleoresins
      c. True Fruit Flavors
   5. Fortified Flavors

Chapter

6. FLAVORS AND FLAVORING—*Continued*   110
   6. Artificial Flavors
      a. Flavor Components
      b. Compounding

7. SPECIALTY AND FRUIT FLAVORS   126
   1. Specialty-Flavored Group
      a. Root Beer and Sarsaparilla
      b. Kola Flavors
      c. Ginger Ale
      d. "Cream" Flavor
      e. Quinine Water
   2. Fruit Flavors
      a. Strawberry Flavor
      b. Raspberry Flavor
      c. Cherry Flavor
      d. Grape Flavor

8. EMULSIONS AND SPECIALTIES   156
   1. Fruit-Flavor Emulsions
   2. Clear-Beverage Citrus-Oil Emulsions
   3. Specialty-Flavor Emulsions
   4. Cloudy-Beverage Citrus-Flavor Emulsions
      a. Physical Properties of Brominated Vegetable Oils
      b. Adjustment of Specific Gravity
      c. Preparation
      d. Formulation
      e. Sugar-Free Formulations
   5. Bodying Agents
      a. Sorbitol
      b. Sodium Carboxymethylcellulose
      c. Other Substances
   6. Foaming Agents

9. COLORS AND COLORING   178
   1. Classification of Colors
   2. Natural Colors
   3. Caramel
      a. Properties

CONTENTS 9

Chapter

9. COLORS AND COLORING—*Continued*   178
    3. Caramel—*Continued*
        b. Manufacture
        c. Use
    4. Synthetic Coloring Matters
        a. Certified Colors
        b. Pure-Dye Percentage
        c. Use

10. CARBON DIOXIDE AND CARBONATION   194
    1. Carbon Dioxide
        a. U.S.P. Requirements
        b. Physiological Response
        c. Preparation
    2. Carbon Dioxide Volume
    3. Determination of Volume
    4. Carbonation
        a. Carbonator
        b. Carbonator and Cooler
        c. Direct Service Carbonation
    5. Formulation
    6. Pressure–Temperature Relationships
    7. Loss of Carbonation

11. BOTTLING AND CANNING   214
    1. Containers
        a. Bottles
        b. Cans
    2. Closures
    3. Bottle Filling
        a. Three-Stage Process
        b. Premix Process
    4. Can Filling
        a. Premix Process
        b. Three-Stage Process
    5. Bottling Fruit Beverages

12. COMPOSITION OF CARBONATED BEVERAGES   231
    1. Sparkling Water, Soda Water, and Seltzer
        a. Sparkling Waters

Chapter

12. COMPOSITION OF CARBONATED BEVERAGES—*Cont.*
    1. Sparkling Water, Soda Water, and Seltzer—*Continued*   231
        b. Carbonated Water
        c. Soda Water
        d. Seltzer
    2. Specialty Beverages
        a. Kola Beverages
        b. Ginger Ale
    3. Fruit-Flavored Beverages

13. PLANT LAYOUT AND SANITATION   243
    1. Plant Layout
    2. Plant Housekeeping
    3. Bottle Washing
        a. Criteria
        b. Washing Machines
        c. Washing Solutions
        d. Predissolving Caustic Soda

14. SPOILAGE   273
    1. Physical Spoilage
        a. Light
        b. Temperature Change
        c. Appearance
        d. Boiling and Gushing
    2. Chemical Spoilage
    3. Biochemical Spoilage
    4. Microbiological Spoilage
        a. Fungi
        b. Yeasts
        c. Molds
        d. Bacteria
        e. Algae
        f. Protozoa

15. CHEMICAL ANALYSIS   288
    1. Water Analysis
        a. Total Solids
        b. Total Hardness

Chapter

15. CHEMICAL ANALYSIS—*Continued*   288
   1. Water Analysis—*Continued*
      c. Total Alkalinity
      d. Available Chlorine
      e. Total Iron
   2. Acid Concentration
      a. Stock Solutions
      b. Acidity in Beverages
   3. Sirup Density and Sugar Concentration
      a. Sirups
      b. Liquid-Sugar Density
      c. Sugar Content of Finished Beverages
   4. Artificial Sweeteners
      a. Saccharin
      b. Cyclamate
   5. Carbonation (Gas Volume) of Finished Beverages
      a. Beverages from Stock
      b. Beverages from Bottling Line
   6. Gas Analysis
      a. Carbon Dioxide and Oxygen
      b. Air in Headspace
   7. Caustic Concentration of Bottle-Washing Solutions
      a. Tablet Methods
      b. Titration Methods
   8. Alkali Concentration of Compounds
   9. Bottle-Rinsing Efficiency
  10. Bottle Uniformity
  11. Crown Crimp Test
  12. Specific Gravity

INDEX   325

Chapter 1

INTRODUCTION

Beverages are foods that are distinguished by two principal characteristics from other foods. First, they are liquids or are consumed in the liquid state and second, they are generally used to quench the thirst.

The major groups of beverages which conform to these characteristics are the carbonated nonalcoholic beverages commonly known as "soda" or "soda pop" and the still beverages, such as fruit drinks and fruit juices. This text will be primarily concerned with the first of these two large groups of beverages.

There is an additional important group of beverages that is characterized by another common property, namely, that of having a stimulating effect, at least initially. There are two principal subgroups in this important group of beverages. One of these comprises the alcoholic beverages and the other the beverages that are made from coffee, tea, and cocoa.

All of the beforementioned beverages have one additional common characteristic, that is, the relative lack of actual food value. In this respect they differ from beverages like milk and other milk products. Indeed the major food value of tea and coffee is the milk and sugar added to these drinks when milk and sugar are used with them.

1. DEFINITIONS

Carbonated nonalcoholic beverages may be defined as beverages that are generally sweetened and flavored, that sometimes are acidified and sometimes have salts or minerals added, that are artificially charged with carbon dioxide, and that contain no alcohol. The name adopted for such beverages by custom, in the United States, is *soda*. This name was derived from the original method of charging the water with carbon dioxide prepared from sodium bicarbonate or sodium carbonate and, as Dr. Harvey W. Wiley [1] predicted years ago, became so firmly associated with "carbonated nonalcoholic beverages" that it has become accepted as the distinctive term for such beverages.

Other terms generally applied to such beverages are *soda pop, soda water,* or more simply, *pop*. Years ago the term "tonic" was also common. In Great Britain, they are known as "aerated waters." As a group such drinks are termed *soft drinks* in contrast to alcoholic beverages.

The term *soft drink,* however, has a broader connotation than merely carbonated nonalcoholic beverage, for the fruit drinks and fruit juices are also included, at times, in the term "soft drink" when they do not contain alcohol.

The group of carbonated beverages can be further subdivided into two groups characterized by their acid content. Acidulated carbonated beverages comprise ginger ales, kolas, and fruit-flavored carbonated beverages. The nonacid type of soft drinks comprise root beer, sarsaparilla, club soda, etc.

In a different type of classification of the carbonated nonalcoholic beverages, three groups are set up. The specialty soft drinks, such as the kolas and root beers, are placed in the first group. The fruit-flavored carbonated beverages are assigned to the second group, and beverages such as club soda, seltzer, etc., are in the third group. The fruit-flavored group can be further subdivided into the citrus-flavored group and other fruit-flavored beverages.

Under the authority of the Wiley Pure Food Act of 1906, the United States Department of Agriculture, Bureau of Chemistry and subsequently the Food and Drug Administration used certain definitions for some of the carbonated nonalcoholic beverages. While such definitions were not reissued under the authority of the Food, Drug, and Cosmetic Act of 1938 by the Department of Agriculture and subsequently the Federal Security Agency, and the Department of Health, Education, and Welfare to which the Food and Drug Administration was transferred, they were considered to have advisory status.

INTRODUCTION

Thus, for instance, the definition for ginger ale was: *Ginger ale* is the carbonated beverage prepared from ginger-ale flavor, harmless organic acid, water, and a sirup of one or more of the following: sugar, invert sugar, dextrose; with or without the addition of caramel color. Ginger-ale flavor or ginger-ale concentrate is the beverage flavor in which ginger is the essential constituent, with or without aromatic and pungent ingredients, citrus oils, fruit juices, and caramel color.

Because of increased knowledge of the toxicity of certain flavoring substances, specifically coumarin and safrole as far as beverages are concerned, it is unlikely that the former definitions for cream soda, sarsaparilla, root beer, and birch beer will be retained. While the former definitions for sarsaparilla, root beer, and birch beer do not mention safrole, oil of sassafras which is mentioned contains 75 to 80 per cent safrole.

In 1954, the Food and Drug Administration published a statement of policy and an interpretation on the "status of foods containing added coumarin." In brief, this statement announced that food containing added coumarin will be regarded as adulterated. In 1957, the Food and Drug Administration issued a call for information concerning the toxicity of safrole. Certain laboratory tests indicated that it had an undesirable degree of toxicity.

The following definition may be adequate for cream soda.

Cream soda or cream soda water, is the carbonated beverage prepared from cream-soda flavor, water, and a sirup of one or more of the following: sugar, invert sugar, dextrose; with or without a harmless organic acid, and with or without the addition of caramel color. Cream-soda flavor or cream-soda concentrate is the beverage flavor prepared from vanilla, vanillin, singly or in combination, together with other flavoring substances; with or without the addition of caramel color.

No definitions have been promulgated for the fruit beverages or fruit drinks either carbonated or still. Some of the carbonated fruit beverages and even some of the noncarbonated fruit-flavored drinks are imitation fruit beverages consisting principally of artificial fruit flavor, artificial color, an organic acid or at times phosphoric acid, sugar, invert sugar, dextrose, and carbonated water. All of these will be discussed in detail.

2. HISTORICAL DEVELOPMENT

a. Mineral-Water Period

The carbonated nonalcoholic beverage industry is considered to have been started by the great British chemist and nonconformist minister Joseph Priestley. It was pointed out by Jacobs [2] in his discussion concerning the historical development of food technology that the manufacture of carbonated beverages was one of the first developments in food technology. Priestley, in 1767, while conducting experiments with carbon dioxide which he collected from beer vats at a nearby brewery, found that the "fixed air," as carbon dioxide was then termed, gave a pleasant acidulated taste to water in which it was dissolved. He also noted that such carbonated waters (or aerated waters in the terminology of his day) resembled naturally occurring carbonated waters. Priestley shortly afterward found a much cheaper way of preparing carbon dioxide by the action of sulfuric acid on chalk. He published his observations in 1772 in a pamphlet entitled, "Directions for Impregnating Water with Fixed Air in order to communicate to it the peculiar Spirit and Virtues of Pyrmont Water, And other Mineral Waters of a similar nature." [3]

While it is usual to associate the origin of the carbonated beverage industry with Priestley, at least in English-speaking countries, one must not forget that naturally occurring carbonated waters were well known long before the eighteenth century and much work of a technological nature had been done on such products. Gübeli-Litscher [4] has covered the history of mineral-water composition and analysis in great detail, going back to Greek and Roman times. Riley,[3] in a preface to a reprint of Priestley's pamphlet, has also stressed this point. As examples one can mention that about 1572, Leonhard Thurneisser (1530–1596), physician of the Elector of Brandenburg, prepared artificial sulfur water, which was then considered useful medicinally. Early in the seventeenth century, ca. 1606, Andreas Libavius (1546–1616) wrote a book, entitled, *De judicio aquarum mineralium* (of the Evaluation of Mineral Waters) on this topic. Even earlier, Paracelsus (1493–1541) noted that gases were formed during fermentation and by the action of acids on chalk.

Priestley was undoubtedly aware of the prior studies and attempts at duplicating naturally occurring effervescent waters made by many

European and British scientists, including van Helmont, Hoffmann, Venel, Hales, Black, Brownrigg, and Cavendish, and many others. He mentions Drs. Macbride, Black, and Brownrigg and Sir John Pringle in his own reports.

Thus Jean Baptiste van Helmont (1577–1644) of Belgium is credited with the discovery of carbon dioxide and with the recognition of the fact that the same gas is formed when wood is burned, acids act on limestone, and certain materials ferment. Johann Joachim Becher (1635–1682) repeated the work of van Helmont. Others who experimented with or prepared artificial mineral waters and carbonated waters were Robert Boyle (1627–1691), the great British chemist, the Swedish chemist Urban Hierne (ca. 1712), Hoffmann, Geoffroy, and Venel. In 1685, Friedrich Hoffmann (1660–1742) physician, chemist, and professor at the University of Halle, suggested the artificial preparation of mineral water by mixing oil of vitriol and alkali in a narrow-mouthed bottle while shaking the vessel to assist in absorbing the gases generated. He did an extensive amount of work on mineral waters, publishing a report entitled, *De Methode Examinande Aquas Salubres,* (of methods for the examination of healthful waters).

Etienne François Geoffroy (1672–1731) attempted to follow the work of Hoffmann, but was not successful. The Frenchman, Gabriel François Venel, around 1750, first impregnated salt-bearing water by generating carbon dioxide with sodium carbonate and acid in a closed container. He thought the gas he had generated was "air" but this was clarified later by Demachy.

In this connection, too, one can mention the work of Stephen Hales (1677–1761) of England, for many of his methods on gases were used later by Black and Cavendish.

Joseph Black (1728–1799), a scientist of Scots ancestry, while professor at Glasgow, did considerable work on carbon dioxide (1754–1757), evolving it from magnesium carbonate (*magnesia alba*) and subsequently from limestone and alkali carbonates. Somewhat later, David Macbride (1726–1778) of Ireland reported his work on gases and similar work was done by Nikolaas Joseph von Jacquin in Vienna about the same time. Some of the work of Black was continued by Richard Bewley who, in 1768, passed carbon dioxide through his preferred salt solutions.

William Brownrigg (1711–1800), an English physician, at about the same time as Priestley, noted that carbon dioxide gave water a pleasant acidulated taste and presented a paper on this topic before the Royal Society at London in 1765. Henry Cavendish (1731–1810), famed for his discovery of hydrogen and for his meticulous and accurate analyses, had developed methods of collecting gases and described these in his report on "Factitious Airs" read before the Royal Society in 1766. In most essentials, the Cavendish and Priestley devices were similar. Cavendish even noted the effect of temperature on the solubility of gases stating that water absorbs more "fixed air" in cold weather than in warm.

Tobern Olof Bergman (1735–1784), a Swedish chemist, who lived at about the same time as Priestley, also made and worked with artificially carbonated water, giving much attention, *ca.* 1770, to the analysis and duplication of mineral waters and their carbonation, generating the gas by use of chalk and oil of vitriol. He proved carbon dioxide dissolved in water to form an acid and called it aerial acid. Much of his work was performed on waters from natural springs. Pierre Joseph Macquer (1718–1784) perfected methods for washing carbon dioxide.

About 1780 Thomas Henry and later his son William Henry, after whom Henry's law is named (see Chapter 10), did their work on the solubility of gases and found that this solubility could be increased by increase in pressure.

Around 1770, the Duke of Chaulnes developed a swirling device for the more rapid and easy mixing of carbon dioxide and water. Some years later, John Mervin Nooth constructed equipment for the saturation of water with carbon dioxide and described this in a paper presented to the Royal Society in 1774. Sulz [5] pointed out that Dr. Nooth's apparatus was identical in principal with the carbonators used around 1880–1890. Gosse and Paul invented an apparatus comprising a pump to increase the gas pressure, a device for washing the carbon dioxide gas, and a stirrer for a mixing chamber. These installations were known for a long time as the Geneva apparatus ("Genfer Apparate").

At the end of the eighteenth century, a number of establishments were manufacturing artificial mineral waters attempting to duplicate well-known spring waters.

b. **Soda-Water Period**

Any thorough search of the literature shows that the line of demarcation between the mineral-water period, in which the principal attempt was to manufacture duplicates of the waters of well-known natural springs, and the soda-water period, in which the principal purpose was to provide a flavorsome, thirst-quenching drink, is not sharp. It also shows that the development of the soda-water industry was simultaneous in a number of countries. Thus in the United States some authorities consider that it was in 1785 that Dr. Philip Syng Physick induced Townsend Speakman, a druggist, to manufacture carbonated waters. This date would correspond roughly with the publication of the original paper of Priestley, while the later date, as noted below, for the start of the industry in the United States, would correspond with the fact that Priestley's life in England was made so uncomfortable, because of his beliefs, that he emigrated to the United States and lived at Northumberland in Pennsylvania. It is possible that Priestley's ideas became better known after he settled in the United States. It may be pointed out that Richard Bewley, an apothecary of Great Massingham, England found, in 1768, that the addition of a small amount of sodium carbonate assisted water in absorbing and retaining increased quantities of carbon dioxide. He was probably the first to make what was long known as "acidulous soda water."

After the publication of Priestley's paper and pamphlet, and this is probably the reason for ascribing so much to Priestley, developments in the preparation of artificially carbonating waters and manufacture on a commercial scale were rapid. John Moncreiff, a druggist and Honor Member of the Royal Society of Edinburg, mentioned in 1778 such preparations and was undoubtedly making such carbonated water in 1794. His book entitled *Medicinal Qualities and Effects of Aerated Alkaline Waters* was published in Baltimore in 1810 and contains descriptions of apparatus for home use.

Chaptal in his book, *Elements of Chemistry*, stated, in 1791, "The carbonic acid is easily dissolved in water. Water impregnated with this acid possesses many valuable medicinal qualities; several apparatus have been successfully invented to facilitate this mixture. The apparatus of Nooth, improved by Parker and Magellan, is one of the most ingenious. On this subject the *Encyclopedie Methodique* may be consulted, article Acide Mephitique."

A most interesting pamphlet is one entitled, "Description of a GLASS APPARATUS, for making mineral waters, like those of Pyrmont, Spa, Seltzer, Etc., in a few minutes, and with very little expense: together with the description of some NEW EUDIOMETERS or instruments for ascertaining the wholesomeness of respirable air; and the method of using these instruments: In a letter to the Rev. Dr. Priestley," by J. H. Magellan, published in

London by W. Parker, 1777, viii + 47 pages with one plate of illustrations. In the 42nd Street main branch of the New York Public Library, this pamphlet is ascribed to Joao Jacinto de Magalhaens. This publication describes in detail a glass apparatus, resembling in some measure a Kipp generator in that it has three parts, a lower section in which carbon dioxide is generated from chalk or marble and oil of vitriol (sulfuric acid), a middle section containing the water to be impregnated and equipped with a valve permitting the carbon dioxide to go up into the middle section but not permitting the water to go down into the bottom section, and a top section having a tube extending halfway down into the middle section. The top section serves to catch the water from the middle section forced up by the gas. The three sections are joined together by ground-glass connections. By use of an additional set of a middle and upper section, one set can be shaken by hand to assist in absorption of the carbon dioxide gas while the other set is being used to impregnate the water it contains.

A glance at this publication shows the advances that were made in only 5 years after the publication of the Priestley directions for impregnating water with carbon dioxide, in 1772.

Other authorities believe that it was not until 1807 that the carbonated beverage industry was started, at least on a commercial basis, in the United States by Dr. Benjamin Silliman (1779–1864), a noted American chemist, in New Haven and Joseph Hawkins in Philadelphia.

In other countries, among the early manufacturers of carbonated waters were Meyer in Stettin in 1787, Paul (as noted before) in Geneva in 1790, and Schweppe in London about the same time. Fries started a business in Regensburg for the manufacture of artificial mineral waters in 1803.

Soda water may have been invented by Augustine Thwaites, an Irish chemist. In 1773, he started importing bottled spring water from the continent. His son, a medical student made a discovery, it is claimed, from which he developed his "Double and Single Soda-Water." This was advertised as early as 1799 in *Faulkner's Journal*. W. F. Hamilton invented and patented a mechanical method of carbonating waters and the Irish rights were purchased by A. L. Thwaites in 1810.

The first patent for the preparation of an imitation mineral water, in the United States, was granted to Joseph Hawkins. He obtained another patent on modifications that he developed in

1823. A patent was granted to Simons and Riondel of Charleston, S. C., in 1810 for a procedure for saturating water with carbon dioxide or fixed air.

There were four major developments in the soft drink industry that contributed to its growth. The first of these was the addition of flavors, such as fruit juices, to the carbonated water. This was probably an American innovation and is attributed to Speakman. In Germany, Dr. Friedrich Adolph Struve, who had set up a series of establishments for the manufacture and sale of artificial mineral waters (Dresden in 1818, Leipzig in 1820, and subsequently in Berlin and other cities), went over to the manufacture of effervescent beverages (Brauselimonaden) which he called "Limonade gazeuse" and which were prepared from pure fruit sirups and carbonated water.

When artificial flavors became available, around 1850, these became common ingredients along with specialty flavors, like ginger ale, root beer, and vanilla extract. The earliest phases of the development of the carbonated beverage industry were closely associated with pharmacies, the products being compounded and dispensed to be consumed directly, that is the beginning of the soda fountain.

The second major development was the manufacture of liquid carbon dioxide. As early as 1823 in England, Sir Humphry Davy (1778–1829) and Michael Faraday (1791–1867) liquefied carbon dioxide. Thilorier, a teacher in the École de Pharmacie in Paris, in 1835, repeated the experiments of Davy and Faraday and produced solid carbon dioxide, but it was not until 1884 that liquid carbon dioxide became available on a commercial basis both in the United States and in Germany.

Liquid carbon dioxide enabled the bottler to eliminate the preparation of carbon dioxide on the premises, a laborious process under the best of circumstances. A related development was the manufacture of solid carbon dioxide on a commercial scale.

The third major development was the crown cap which was invented by William Painter in 1892. Some idea of the importance of this invention can be gathered from the fact that over 1500 different devices had been used and suggested for use for the stoppering and closing of bottles for carbonated beverages before this development.

The fourth major development was the introduction of machine-made bottles. In the United States, the successful bottling of the famous Saratoga Springs, N. Y., mineral waters was followed by the bottling of flavored soft drinks. The uniformity of machine-made bottles permitted the development of many other machines, such as filling and capping machines and thus assisted materially in the growth of the industry.

A picture of the growth of the specialty-flavor market can be visualized from the exotic names that were common in the 1882–1884 period. Among these were "Little Daisy," "Inca Cocoa," "Buffalo Mead," "Imperial Nerve Tonic," "Fillipin Fizz," "Klondike Fizz," "Kolafra," "Ironbrew," "Kola Phosphate," "Tangerette," "The Dutchess," "Bongo Beer," "Grapine," "Bull Ginger Ale," "Ho-Ko," "Lime Phosfizz," and "Cycla-Phate."

3. PRODUCTION AND SALES

The soft drink industry of the United States is a billion-dollar industry. In 1957, it was estimated that sales amounted to $1,300 million, that some 5,200 establishments were engaged in bottling carbonated beverages, and that these establishments employed nearly 100,000 persons. The industry, itself, is a large consumer of materials. These data are shown in Table 1-1.

TABLE 1-1

Sales and Materials Used in the Carbonated Beverage Industry [a]
1957

Item	Value or amount
Sales	$1,300,000,000
Establishments	5,400
Employment, persons	100,000
Diverse outlets	1,500,000
Cooling and vending machines	2,000,000
Sugar, tons	1,500,000
Bottles, gross	9,000,000
Crowns, gross	200,000,000
Delivery cases	17,000,000
Cans, gross	2,500,000
Gasoline, gallons	100,000,000
Trucks	60,000
Advertising	$100,000,000

[a] After *Natl. Bottlers' Gaz.*, **76**, No. 901 (March, 1957).

The growth of the industry in this country is interesting. The earliest records show that the carbonated-beverage industry was well recognized by 1835. In the 1850 census, there were 64 establishments with a capital of $228,650, having a production value of $760,000. By 1875, there were some 512 establishments having a production of $4,740,000. The invention of the crown cap, the production of machine-made bottles, and the availability of liquid carbon dioxide, as mentioned, enabled the industry to expand in the twentieth century. Its development in recent years is shown in Table 1-2.

TABLE 1-2

Bottled Soft Drinks
General Statistics for the United States, 1947–56

Census year	Establish- ments	Employees, total		Value added by manufacturer ($1000)	Cost of materials ($1000)	Value of shipments ($1000)
		No.	payroll ($1000)			
1956[a]	5,200					1,308,000
1954	4,643	91,616	319,879	635,045	478,034	1,113,077
1953		99,106	339,509	687,036	497,882	1,184,918
1952		83,939	266,141	541,040	393,060	934,100
1951		84,189	242,621	508,553	368,418	876,971
1947	5,618	79,397	194,305	420,711	327,485	748,196

[a] Am. Bottlers Carbonated Beverages estimate.

One may note in comparison that in the 1954 census of manufacturing, issued by the United States Department of Commerce, the total value of shipments for the carbonated beverage industry was $1,113 million of which $1,087 million were primary products or bottled soft drinks, $18 million accounted for secondary products, and $8 million of miscellaneous receipts. Some $76 million worth of products were bought and sold and were not included in the total. In the beer and ale industry, for 1954, total shipments of primary products amounted to $1,851 million; there were $4 million of secondary products; and $2 million of miscellaneous receipts. In the distilled liquors industry, total shipments of primary products in 1954 amounted to $683 million; there were $8 million of secondary products; and $21 million of miscellaneous receipts.

Most of the carbonated soft drinks are made by franchised bottlers who obtain their flavored sirups or flavors from large organizations. Some state that such bottlers manufacture 90 per cent of the carbonated beverages. About 50 to 60 per cent of such bottled drinks are sold by food stores. This percentage doubled over the amount sold in such stores in the 1940's. This big increase may be due to the growth of the supermarket industry in this period with the open display of the soft drinks spurring their purchase and possibly also to the influence of television with more persons staying at home and thus consuming bottled beverages at home rather than at the fountain.

It has been estimated that sales from vending machines account for 15 to 20 per cent with the remainder being sold by candy stores, drug stores, delicatessens, service stations, and restaurants.

One significant step forward in the 1950's has been the development of premix vending machines. Another has been the use of cans for soft drinks. Both of these will be discussed in later chapters.

The earnings of the largest carbonated beverage organizations indicate that the most popular flavor is kola. The national preference is kola, orange, root beer, ginger ale, lemon, and lime. In Table 1-3 the regional preferences are tabulated. There is, of

TABLE 1-3

Flavor Preferences for Carbonated Beverages [a]
(percentage)

Flavor	Region of the United States			
	South	West	North Central	Northeast
Kola	61	38	34	26
Ginger ale	4	4	5	23
Root beer	7	22	17	11
Orange	10	16	18	13
Lemon and lime	2	10	10	10

[a] After *Natl. Bottlers' Gaz.*, **74**, No. 883, 46 (1955).

course, a moderate consumption of fruit flavors. Surveys have shown that some three out of ten families drink sparkling beverages with meals, that is 29 per cent of the families in the entire country do this. This is the practice of 42 per cent of the families in the South, 32 per cent in the Northeast, and 20 per cent in both the West and the North Central States. Many consumers of soft drinks use them as mixers for alcoholic beverages. In the United States as a whole, 44 per cent followed this practice. The regional percentages are West 55 per cent, South 45 per cent, North Central States 42 per cent, and Northeast 40 per cent. A survey conducted by *American Girl* in 1957 on the flavor preferences of teen-age girls (10 to 16 years of age) showed that 40 per cent preferred kola, 26 per cent citrus, 15 per cent root beer, 7 per cent grape, and 4 per cent ginger ale. These girls consumed on the average 4.4 split-size bottles of soft drinks a week totaling about 228 per year and, in addition, drank about $1\frac{1}{2}$ glasses from large size bottles each week. Drinks like ginger ale and club soda which are used for mixers with alcoholic beverages have greater sale to adults.

LITERATURE CITED

1. H. W. Wiley, *Beverages and their Adulteration*. Philadelphia, P. Blakiston's Son & Co., 1919.
2. M. B. Jacobs, ed., *Chemistry and Technology of Food and Food Products*, 2nd Ed., New York, Interscience Publishers, 1951.
3. Reprint with an introduction by J. J. Riley, Washington, Am. Bottlers Carbonated Beverages, 1945.
4. O. Gübeli-Litscher, *Chemische Untersuchung von Mineralwässern*, Innsbruck, Universitätsverlag Wagner, 1948.
5. C. H. Sulz, *Treatise on Beverages*, New York, Dick & Fitzgerald, 1888.

SELECTED BIBLIOGRAPHY

J. J. Riley, *Organization in the Soft Drink Industry*, Washington, Am. Bottlers Carbonated Beverages, 1946.
G. H. Dubelle, *Soda Fountain Beverages*, New York, Spon & Chamberlain, 1892.
E. Walter, *Manual for the Essence Industry*, New York, Wiley, 1916.
R. H. Morgan, *Beverage Manufacture*, London, Attwood & Co., Ltd., 1938.
R. Kühles, *Handbuch der Mineralwasser-Industrie*, Lübeck, Antäus-Verlag, 1947.
Natl. Bottlers' Gaz., **76**, (75th Anniversary issue) (1957).

Chapter 2

SUGARS AND SIRUPS

The manufacture of carbonated nonalcoholic beverages appears at first glance to be deceptively simple as compared with the manufacture of certain prepared foods or with other types of chemical processing, but closer examination, as will be clear from this book, shows that many of the steps, while not complex—with the exception of those connected with the preparation of the flavoring—require great care. As is true of all types of food manufacture, particular care must be taken to have proper plant housekeeping and complete sanitary control. These phases will be discussed in detail in Chapter 13.

Generally the process of manufacturing carbonated soft drinks comprises the preparation of a sirup from water and sugars or more concentrated sugar sirups; adding acidulants and flavorings; blending the mixture of ingredients; transferring definite volumes of the blended mixtures to bottles or cans; adding sufficient carbonated water; capping or sealing the cans, and labeling; and then distributing the containers. A flow chart of the various steps used in the manufacture of carbonated soft drinks is shown in Figure 1.

1. SUGARS

As noted in the definitions of some of the more important carbonated soft drinks given in Chapter 1, the principal sugars and sirups used in the manufacture of these soft drinks are sucrose, dex-

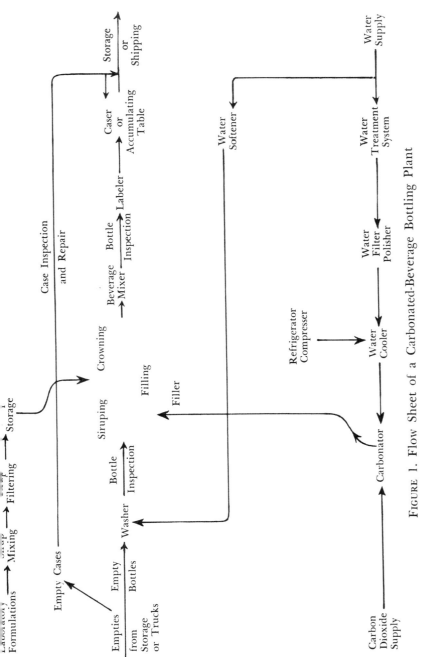

FIGURE 1. Flow Sheet of a Carbonated-Beverage Bottling Plant

trose (D-glucose), and liquid sugar. Other sugars, such as fructose (levulose), lactose, maltose, and the like, are rarely used except in times when common sugar is not readily available for they are more expensive.

Sirups other than liquid sugar as, for example, liquid corn sugar (not to be confused with commercial glucose), invert sugar, honey, and cereal sirups are also seldom used for carbonated beverage manufacture except in times of severe common-sugar stringency.

The sugar and sirup ingredient of carbonated nonalcoholic beverages serves, as has been stressed by Buchanan [1] and many others, as more than a mere sweetening agent. It has three principal functions, namely, (1) it provides the sweetness to balance properly the acid and other taste-producing components of the beverage and thus produces a balanced flavored soft drink; (2) it furnishes sufficient body to raise the beverage out of the watery class; and (3) it serves to carry the flavor and thus deposit it uniformily when consumed.

The major food value of carbonated beverages is obtained from the sugar ingredients. The over-all amount of caloric value is not large as can be seen from the following values:

Sugar, % Brix	Calories 8-oz. bottle
11.0	108.4
12.5	123.9
14.0	139.6

Thus, since most carbonated beverages contain less than 10 per cent of sugar, the total amount of calories contributed by an 8-ounce bottle is less than 100 calories.

a. **Sucrose**

Sucrose, $C_{12}H_{22}O_{11}$, a disaccharide, is a sugar that is characteristic of plant materials. It is obtained on a commercial scale from sugar cane and sugar beet. It may also be obtained from sorghum, sugar maple, sugar palm, and numerous other plants. For example, about 50 per cent of the dry matter of some vegetables like the carrot and some fruits like pineapple is sucrose.

Common sugar or sucrose is the principal sugar used in the carbonated beverage industry. In 1952, an estimated 1 million short tons or about 12.5 per cent of the 8.1 million short tons of sucrose used in the United States was utilized by the carbonated-beverage industry.

Sugar used to be delivered to the industry in barrels or in cotton or burlap bags. The first type of packaging was satisfactory, but the second type is not suitable for if care is not taken, sugar packed in cotton or burlap bags easily becomes contaminated with dirt and dust. In more recent years, 100-pound capacity paper bags have been found to be adequate. Large bottlers receive bulk shipments of sucrose in carload and truck lots. The bulk sugar is handled by screw conveyors, batched in bins, and moved by forked trucks.

(1) Bottlers' Tentative Sugar Standard

In a bulletin issued in 1953, American Bottlers of Carbonated Beverages [2] pointed out that for many years bottlers of carbonated beverages had been asking for a sugar standard that would aid them in purchasing granulated sugar that would consistently meet the exacting requirements for the production of quality beverages. It was stressed that many times, excessive impurities present in sugar have been the cause of foaming at the filler, resulting in shortfilled bottles, beverage ingredients spilled on the bottle surface and reduction in the speed of operation; sediment in the beverage; off-taste or masking of the beverage flavor; off-odor or yeast or bacterial spoilage. It was noted that the tentative standard was not a guarantee that all the conditions mentioned would be eliminated, but it was concluded that the purchase of sugar meeting the requirements would go a long way in obtaining uniform production and in the improvement of the quality of bottled carbonated soft drinks. They urged that all beverage manufacturers purchase and accept only sugar in packages that are stamped "Bottlers' " designating that the sugar has been tested and conforms to the industry standard.

(a) Sampling

The following precautions are recommended in the taking of samples for submission to a laboratory for analysis or for analysis by the laboratory of the firm purchasing the sugar.

(1) Container—one-pint glass jar having a glass lid without rubber gasket. Detachable metal lids may also be used; the lid, however, should be inverted to avoid the possibility of the sugar absorbing or adsorbing the odor from the rubber gasket. Before

collecting the samples, the jars must be cleaned and then sterilized by heating in an oven at 329°F for 1½ hours.

(2) Take sugar samples only from freshly opened sugar bags. To avoid the possibility of including in the sample any dust that may have fallen on the outside layer of the sugar during the opening of the bag, remove the top layer of the sugar before taking the sample.

(3) Fill the pint jar to within 1 inch of the top to provide sufficient sample for all analyses.

(4) Include on the label of the jar, the brand name of the manufacturer, the type of container in which the sugar was packed, and the code mark on the bag or container.

(b) Tentative Standard Requirements

This tentative standard applies to dry granulated sugar only and is not applicable to liquid sugars. These standards apply to sugar as produced, immediately prior to packing.

Container—For the protection of the product, "Bottlers'" sugar shall not be packed in cotton or fabric bags, but shall be packed in multiwall paper bags or equivalent sanitary packages or bulk containers.

Container identification—Each container shall be marked or coded to make it possible for the sugar producer to identify the place of production and date of packing.

Designation of type—Each container shall be marked "Bottlers'."

Ash—The ash content of "Bottlers'" sugar shall not be more than 0.015 per cent.

Color—The solution color of "Bottlers'" sugar shall not be more than 35 reference basis units.

Sediment—The sediment content of "Bottlers'" sugar shall not be more than shown on a prepared sediment disc available from American Bottlers of Carbonated Beverages.

Taste and odor—"Bottlers'" sugar shall have no obviously objectionable taste or odor in either dry form or in a 10 per cent sugar solution prepared with tasteless, odorless water.

Bacteriological—"Bottlers'" sugar shall not contain more than:

        200 mesophilic bacteria   per 10 grams
        10 yeast   per 10 grams
        10 mold   per 10 grams

Sampling—"Bottlers' " sugar shall be adequately sampled by the producer just before packing to assure compliance by test to these standards.

There are two other aspects of quality which are of great importance to the bottling industry, namely, (1) turbidity and (2) floc-producing substances. Universally accepted methods of testing for these factors have not yet been developed. For this reason the foregoing standards do not cover turbidity and floc-producing substances. As soon as accepted test methods and tolerances have been established, it is expected that they will be added to the standard. It is imperative that "Bottlers' " sugar is as nearly free of turbidity and floc-producing substances as possible.

Reference basis units for both ash and color have been developed by the joint research of the carbonated beverage and sugar industries and thus make possible the expression of test results on a common basis. Information concerning such units is available from American Bottlers of Carbonated Beverages.

b. Liquid Sugar

Within the past two decades, the use of "liquid sugar," has become common in the carbonated-beverage industry. Such products can be made to have almost the same purity as crystalline sucrose. There are two major types of liquid sugar, namely, 100 per cent sucrose No. 1 liquid sugar and 50/50 invert-sucrose sirup. The No. 1 sucrose sirup is made by two principal methods. One of these consists of boiling down No. 1 sugar liquors that have been treated with bone char and vegetable carbon. In this type of manufacture, the purified sucrose solutions are filtered through char. All the taste and odor are removed by additional purification with the aid of vegetable carbon, after which the solution is concentrated to 68° Brix. In the second method, granulated sugar is dissolved in water and the sirup formed is subsequently sterilized and filtered. The dissolved granulated sugar product will have a low ash content and will comply with standards for granulated sugar.

Gortatowsky[3] has pointed out that liquid sugar in which the sucrose has not been boiled out in a pan will have a higher ash content. Products made by both methods are utilized by the beverage industry. Liquid sugars of higher ash content may have 0.05 per cent ash on a wet basis. Some beverage manufacturers find such an ash content objectionable. Other manufacturers feel that the nonsugar organic content is of greater importance, particularly if this nonsugar portion contains acids of the fatty acid series. Such organic acids may be set free from their salts on acidification of the beverage sirup and give rise to objectionable odors.

Liquid sugar lends itself to ease of handling at the plant. The concentrated sugar solution may be pumped directly to a storage tank and from there to the sirup tanks. Liquid sugar is generally sold at a lower price than the corresponding amount of crystalline sucrose since it has required less processing. When it is necessary to transport it over long distances, however, the additional water that has to be carried reduces the economies that can be expected by its substitution for sucrose.

The liquid sugars on the market may be classified as follows:

(1) A liquid sucrose solution, made as noted before, that contains about 67 per cent of sucrose and is water-white. The analysis of a typical No. 1 sucrose liquid sugar is 67.0 per cent solids, 0.2 per cent invert sugar, 0.05 per cent ash, pH 6.8, water white, and showing no turbidity.

(2) A liquid sucrose solution also containing about 67 per cent of sucrose, but leaving more ash than the first category and having a slight coloration.

(3) A partially inverted liquid sugar that contains about equal quantities of sucrose and invert sugar (often termed 50 per cent invert or 50/50 invert-sucrose), yielding a mixture equivalent to about 78 per cent of sucrose in sweetening power attributable to the greater solubility of sugar in the mixture. There are two principal reasons for the belief that this type of liquid sugar is more suitable for the manufacture of carbonated beverages. The first reason is that since it contains less water, the manufacturer is not paying for the shipping cost of so much water as with straight sucrose liquid sugar. The second reason is that, since the solids content is higher, there is less chance for yeast contamination. The highest grade of 50 per cent invert liquid sugar is water white. A typical analysis of a 50 per cent invert sugar is 76 per cent solids, 0.05 per cent ash—or as low as 0.01 per cent ash if the product is made by the ion-exchange method, pH 5.0, straw color that develops through inversion, and no turbidity.

(4) A partially inverted liquid sugar, containing about equal quantities of sucrose and invert sugar but more ash and color than the highest grade liquid invert sugars.

All of these liquid sugars should be free from yeast, bacteria, and molds.

(1) Bottlers' Tentative Liquid Sugar Standard

In a bulletin published in 1956, American Bottlers of Carbonated Beverages [4] issued a tentative standard for liquid sugar. It was noted, as in the case of granulated sugar, that the tentative standard is not a guarantee that all problems that may be associated with the use of liquid sugar will be eliminated, but bottlers who purchase liquid sugar that meets this industry standard will go a long way toward the production of soft drinks of uniformly high quality required to compete with other beverages on the market. They urged that all beverage manufacturers who use liquid sugar inform their suppliers that they require liquid sugar meeting the industry standard.

Shipping Unit—"Bottlers' " liquid sugar shall be shipped in clean, sanitary shipping units. The shipping unit shall be so constructed as to eliminate any possibility of adding any impurity to the liquid sugar.

Identification—The product shipped shall be identified on the bill of lading as "Bottlers' " liquid sugar and shall specify whether it is sucrose type or 50 per cent invert type.

Color—The solution color of "Bottlers' " liquid sugar shall be the same as that established for "Bottlers' " granulated sugar—not more than 35 reference units.

Sediment—The sediment content of "Bottlers' " liquid sugar shall be the same as that established for "Bottlers' " granulated sugar.

Taste and Odor—"Bottlers' " liquid sugar shall have no objectionable taste or odor in an undiluted form or in a 10 per cent solution acidified to pH 2.5 with U.S.P. phosphoric acid.

Copper, Iron, Zinc—"Bottlers' " liquid sugar shall not contain more than 1.5 ppm iron, copper, or zinc.

Bacteriological—"Bottlers' " liquid sugar shall not contain more than:

100 mesophilic bacteria	per 5 grams
2 yeast	per 5 grams
5 mold	per 5 grams

Sampling—"Bottlers' " liquid sugar shall be adequately sampled by the producer, to assure compliance with these standards. Conditions of liquid-sugar storage tanks, lines, etc., in the bottling plant is the responsibility of the bottler.

Turbidity and floc-producing substances—The comments noted under the standard for granulated sugar apply to the standard for liquid sugar also.

c. Invert Sugar and Sirups

Invert sugar is a mixture consisting of approximately 50 per cent of dextrose (D-glucose) and 50 per cent fructose. It is made by the hydrolysis of sucrose solutions. As mentioned in the previous section, one type of liquid sugar contains about 50 per cent of invert sugar.

A number of invert sirups are made in metropolitan regions for use in products such as beverages and confectionery and lend themselves to ready distribution in such areas. They generally contain as detailed in the preceding section about 76 per cent of solids about half of which is invert sugar and the other half sucrose. They also contain small amounts of ash and color which may range from water white through straw-yellow to golden yellow.

A type of invert sugar not commonly used in the beverage industry, but which might find some use in times of sugar stringency, is sold in a semiplastic form. It is prepared by the inversion of a very concentrated sucrose solution by hydrochloric acid at a pH of 3 to 4 with the aid of the enzyme, invertase. The acid is neutralized to a pH of about 6 with sodium carbonate and, when the crystallization of the dextrose reaches equilibrium, the entire mass is beaten into a creamy product.

d. Dextrose Sugars

Dextrose or corn sugar whose proper chemical name is D-glucose, is commercially available in a number of forms, namely, crude dextrose sugar, pressed dextrose sugar, cerelose, anhydrous dextrose, and dextrose hydrate. Very great care has to be taken in the manufacture of dextrose sugars. The precautions have been described in detail by Newkirk.[5]

Dextrose sugars have found some utilization in the carbonated-beverage industry because they can be used to give body to a beverage without added sweetness. It is well known that dextrose does not have as great sweetening power as an equal weight of sucrose. Thus, if the sweetening power of sucrose is taken as 100, then the sweetening power of D-glucose is about 74. Buchanan [1] has pointed out how advantage can be taken of this difference in sweetening power to improve flavor. For instance, a grape beverage which had been manufactured with a concentration of about 14 per cent of sucrose was found to be too sweet. The mere addition of acid could not alter this overpowering sweetness and only made the beverage too acid. By using sufficient dextrose to make up 30 to 40 per cent of the total sugar content, the sweetness was reduced sufficiently without any reduction in the desired body of the beverage. It seldom occurs, however, that dextrose is used to replace more than 20 per cent of the customary sucrose concentration.

About 45,000 to 50,000 tons of dextrose are used annually by the carbonated-beverage industry which is a relatively small proportion of the million tons of sucrose used by the industry. Gortatowsky [3] noted that fruit drinks, particularly grape-flavored drinks, and root-beer beverages were the relatively more important soft drinks in which dextrose was used as a sugar ingredient. He pointed out that a larger percentage of bottlers of root-beer soft drinks used dextrose in comparison to bottlers of fruit-flavored beverages. Thus 33 per cent of the root-beer bottlers, 20 per cent of the lemon-lime bottlers, and 18 per cent of the ginger-ale bottlers used dextrose in concentrations ranging from 5 to 45 per cent of the total amount of sweetening agent used. The greater number of bottlers preferred to limit the concentration of dextrose to the range of 12 to 25 per cent. In his survey he found that only one bottler used corn sirup for the manufacture of a carbonated beverage.

(1) Crude Dextrose Sugar

The crude residue from the vacuum pans, in the manufacture of dextrose from corn, is poured into tanks where partial crystallization is permitted to take place. The mass is cut into slabs and aged so that additional crystallization may take place. Such products are known as "70" or "80" sugar, depending on the approximate percentage of dextrose present. The dextrins that these sugars contain give the crude dextrose sugar a light-brown color.

(2) Pressed Dextrose Sugar

Pressed dextrose sugars are obtained by squeezing the "80" type dextrose sugar coming from the vacuum pan in hydraulic presses. This treatment separates most of the crystalline sugar from the dark-colored mother liquor, which is termed corn molasses or hydrol.

The resulting cakes are ground and dried, giving a product that is higher in D-glucose content and lighter in color than the starting material.

(3) Cerelose

The heavy sirup from the vacuum pan obtained by evaporation of almost completely converted starch liquors is permitted to stand in crystallizers, similar to those used in the manufacture of sucrose, at about 90°F after being grained with pure dextrose. After crystallization, the mother liquor is removed by centrifugal methods analogous to those used for separating cane sugar. The cerelose obtained in this manner contains about 91 per cent of D-glucose, the remainder being virtually all water.

(4) Anhydrous Dextrose

Anhydrous dextrose is made from cerelose by special refining methods, using equipment made of copper, brass, bronze, and Monel metal. The final product contains 99.5 to 99.8 per cent of D-glucose.

2. BEVERAGE-SIRUP MANUFACTURE

Having selected the sugar or sirup most suitable for his purpose, the carbonated-beverage manufacturer employs five major methods for the preparation of his beverage sirup. The principal objective of all of these methods is to obtain a sirup of the proper sugar concentration that is free from microbiological or other contamination. Because the final beverage is not heat treated and also because the beverage sirup is an excellent medium for the growth of many bacteria, yeasts, and fungi, the greatest care must be taken to use uncontaminated raw materials and prevent subsequent contamination of the sirup. The general and special aspects of beverage-plant sanitation are considered in Chapter 13.

The five major types of sirup used by the soft-drink manufacturer are (1) cold-process sirup, (2) hot-process sirup, (3) acidified cold-process sirup, (4) acidified hot-process sirup, and (5) high-density sirup.

a. Simple Sirups

The solution obtained by dissolving sugar in water is known as simple sirup. The definition of the *United States Pharmacopeia* for simple sirup is a sirup that contains 850 grams of sucrose per liter of sirup. This is equivalent to 36°Bé. This definition is in contrast to the simple sirups used in the manufacture of carbonated nonalcoholic beverages. Such sirups contain 48 to 59 per cent of sugar by weight, equivalent to 4.88 to 6.30 pounds of sugar per gallon, corresponding to sirups of 26° to 32°Bé. The variation is due to the individual formulations of the beverage manufacturer, but most bottlers use 30° to 31°Bé sirups, because such sirups flow freely at the temperatures existing in most bottling plants.

b. **Cold-Process Sirup**

There are several variations in the manufacture of cold-process sirup, depending on whether solid sugars or liquid sirups are used. Basically, the cold process of sirup preparation consists of mixing and dissolving sugar or liquid sugar or sirup in water at room temperature. It has several advantages for less expensive equipment is necessary and the cost of heating and subsequent cooling is eliminated. However, since the sirup being prepared at lower temperatures is more viscous than that prepared at higher temperatures, mixers and stirrers of greater horsepower are required to prepare the solutions. Moreover, since no heating is used, in effect a pasteurization step is omitted, and exceptional care must be taken to avoid contamination of the sirup.

In preparing cold-process sirup with crystalline sugar, treated water—made as detailed in Chapter 5—is passed into a vessel termed a mixing tank. This vessel is fitted with one or two stirrers, depending on its size. The agitation is started and then the crystalline sugar is transferred into the tank from storage. The mixing should be adequate to assure full solution, but should not be too vigorous for proper mixing avoids the incorporation of air and possible contamination from the air.

The water used is generally not heated or warmed but, if dextrose is being used, sufficient hot water is first added to dissolve the dextrose and then enough unheated water is poured in to dissolve the sucrose that is added last. The entire mixture is stirred until all the sugars are dissolved.

The mixing vessels should be made of stainless steel (usually 18-8, type 304 number of finish), or of an alloy of nickel, or be glass-lined. Virtually all of the vessels used in the carbonated beverage industry are made of stainless steel. A few glass-lined vessels are used. The vessels should be equipped with sanitary piping, pumps, and valves. There should be a shut-off valve at the drain port of the tank. This can be used to avoid clogging the drain with undissolved sugar or sirup. The stirrer will require a motor of about $\frac{1}{3}$ to $\frac{1}{2}$ horsepower with a tank of 500-gallon capacity.

As mentioned sugar can be delivered in bulk by carload or truck lots. This sugar can be handled by bulk methods for the preparation of sirup as, for instance, by the use of a screw conveyor or tote bin.

Various methods of making cold-process sirup on a large scale, equivalent to the delivery of liquid sucrose or sugar, have been developed. One of these methods makes use of a device termed a "Liquilizer." A portable 25-pound, 3-foot long metal converter or dissolving device [6] is attached to the outlet flange of any railroad car or truck used to transport and deliver bulk quantities of granulated sugar. This device mixes the sugar and water according to the purchasers' needs while the sugar is being unloaded at the plant. The sugar falls down into the dissolving device and mixes with water fed through stainless-steel pipes from the plant, itself. The blended mixture is pumped into the plant's mixing tank where it is stirred thoroughly. The water used can be measured accurately and in accordance with the amount of dry sugar being delivered. This device can unload and blend 140,000 pounds of bulk sugar in 5 hours and provide a cold-process sirup ready for use.

If liquid sugar is being used to make the cold-process beverage sirup, the preparation is simpler. Transfer the calculated amount of water to the mixing tank and pump the required amount of liquid sugar to the tank. Mix by stirring slowly until solution is complete.

In both instances, determine the sugar concentration of the sirups by taking hydrometer readings using either a Brix hydrometer, which gives readings directly in sugar percentage, or a Baumé hydrometer. In the second case, a table will assist in converting to sugar percentage.

The location of the sirup preparation tanks is important. Generally, if dry sugar is used, the mixing tank is located in the sirup room even if it is on the second floor. If a mixing tank is being used for the preparation of cold-process sirup from crystalline sugar, it is preferable to locate the tank on the ground floor. This enables the operator to prepare the sirup near the storage of the crystalline sugar and avoids the transportation of the solid sugar farther than necessary. The beverage sirup can then be pumped to the mixing tanks usually located in a sirup room on a balcony.

If liquid sugar is being used, the initial mixing tank can be located on the same floor as the mixing tanks of the sirup room.

It is important to remember that if the sirup made by the cold process is not to be used immediately or shortly after preparation, particularly if the sirup is light to medium, that is, 30° to 34°Bé or less, it should be acidified for preservation and even then should not be held over longer than a weekend. Another important point to remember is that if a prepared flavor, that contains all of the acid necessary for the beverage, is being used by the beverage manufacturer, then only sufficient beverage sirup should be made for the amount of flavor to be used. There should be no carry-over of sirup. Sirups with specific gravities of less than 36°Bé should not be stored, for yeasts are capable of growing rapidly at such relatively low sugar concentrations. Under certain conditions, yeasts will grow rapidly even in 36°Bé sirup.

It is also important to remember that when cold-process sirups are used for the preparation of acid-type carbonated beverages, some of the sucrose is inverted on standing. The amount of inversion is dependent on the length of storage. An example of this action is shown in Table 2-1. It can be seen that in this experiment nearly 50 per cent of the sucrose was inverted in about 2 months.

TABLE 2-1

Inversion of Sucrose,[a]
Effect of Time

Days	Sucrose %	Reducing sugar, %
0	8.37	0.53
1	8.35	0.55
4	8.10	0.80
8	7.70	1.20
17	7.10	1.80
32	6.10	2.80
46	5.33	3.57
59	4.48	4.42
112	2.80	6.10

[a] After J. H. Buchanan, in *Bottlers' and Glass Packers' Handbook*, p. 51, New York, Food Engineering, McGraw-Hill.

Gortatowsky also has commented on the inversion of sucrose in acid-type beverages, which constitute the bulk of those sold. Thus in beverages having a pH in the range of 2.5 to 3.5, the sucrose will undergo inversion and this will be complete particularly if left to stand at a high enough temperature for a sufficient length of time. His analyses show that in beverages having a pH of 2.5, about 50 per cent of the sucrose is inverted in about 25 days at room temperature and specimens of the same beverages were completely inverted at the end of 230 days. It may thus be inferred that a carbonated beverage belonging to the acid-type group will have when consumed more than 50 per cent sucrose and less than 50 per cent invert sugar.

The flavor aspects of this change must be taken into consideration in the preparation of the beverage, for invert sugar has a flavor that is different from that of sucrose and this must be taken into account in evaluating the flavor of the final beverage.

c. **Acidified Cold-Process Sirup**

The product obtained when acid is added to simple sirup prepared by the cold process is known as *acidified simple sirup*. When all of the flavoring components and ingredients have been added and mixed together, the resultant product is known as *flavored sirup*.

Acidified cold-process sirup is relatively less subject to attack by microorganisms than unacidified simple sirup. Table 2-2 shows the effect of acid in preserving simple sirup in an experiment. It must be stressed, however, that yeasts vary in their ability to get acclimated to different conditions of environment and thus the data in Table 2-2 are not to be considered as absolute.

TABLE 2-2

Days Required to Produce Sterility [a,b]

Sirup Bé	Citric acid, 50% per gallon sirup		
	1 oz	2 oz	4 oz
24	5	4–5	3–4
30	2–3	2	1–1½
36	1–1¼	1	1

[a] After J. H. Buchanan, in *Bottlers' and Glass Packers' Handbook*, p. 51, New York, Food Engineering, McGraw-Hill.
[b] See discussion in text.

SUGARS AND SIRUPS

Acidified cold-process sirup will undergo inversion in a manner similar to that noted in the preceding section.

In preparing acidified cold-process sirup, it is important to add the acid slowly and to stir slowly before the flavoring is added. In preparing the flavored sirup, this need for slow mixing and stirring is of even greater importance for this will avoid the incorporation of air which will, in turn, avoid oxidation of the more readily oxidizable flavoring components. There are instances in which too rapid mixing and the incorporation of air has caused spoilage in the manufacture of citrus-flavored beverages, such as orange.

d. Hot-Process Sirup

In the production of hot-process sirups, the water and sugar are heated to facilitate the solution of the sugar. An additional objective is to kill microorganisms. There are two general submethods of preparing hot-process sirups. The first is a boiling method and the second is in reality pasteurization. Such sirups are preferred if they are to be stored for several days before the final beverage is manufactured.

In the boiling method, 32°Bé sirup, usually prepared in steam-jacketed, stainless-steel, 50- to 200-gallon vessels equipped with sanitary fittings, pipes, stirrer, valves, and pump, is heated to boiling and held at that temperature for 5 minutes. It is then cooled by circulating cold water through the vessel or preferably, since this is a slow procedure, by passing the hot sirup through a filter press or other filter device and then to a cooler or heat exchanger.

In the pasteurization method, 32°Bé sirup, containing approximately 1 per cent acidulant is heated to 180°F for a given period of time, and is then filtered and cooled as noted previously.

In another variation of the pasteurization process, the sugar is added to hot or cold water with constant stirring and steam or hot water is circulated in the outer jacket until the temperature of the thoroughly mixed sirup reaches 190° to 200°F. The sirup is held at that temperature for 10 minutes and is then cooled rapidly. This method should produce a completely sterile product.

Neither process guarantees that the sirup will be free from contamination after processing. It is essential that adequate handling facilities and sanitary storage are available for all subsequent beverage-manufacturing operations.

e. **Acidified Hot-Process Sirup**

Acidified hot-process sirup is prepared by the addition of acid before or during the heating step. This results in the inversion of the sirup. Since the sirup is virtually completely inverted, there is little, if any, additional flavor change on standing. Actually, the pasteurized variety of the hot-process sirup detailed in the preceding section is an acidified hot-process sirup. Table 2-3 shows that inversion takes place rapidly at 180°F.

TABLE 2-3

Sucrose Inversion
Effect of Temperature [a] and Acid [b]

Sp gr Bé	Citric acid, 50% oz/gal	Inversion %
24	1	94.44
24	2	97.44
30	1	97.10
30	2	97.80
30	4	98.70
33	2	97.40
33	4	98.40

[a] Thirty minutes at 180°F.
[b] After J. H. Buchanan, in *Bottlers' and Glass Packers' Handbook*, p. 51, New York, Food Engineering, McGraw-Hill.

f. **High-Density Sirup**

As mentioned in a previous section, many simple sirups used by beverage manufacturers range from 26° to 32°Bé, corresponding to 48 to 59 per cent of sugar by weight. In addition to the hot process for the prevention of microbial growth in sirups, another method is available to the manufacturer, namely, that of using high-density sirups. It has been shown that sirups which have a density of 36°Bé, that is, sirups with a sugar concentration of 67 per cent of sugar equivalent to about 7.4 pounds of sucrose per gallon, retard the growth of microorganisms. Thus sirups that have to be stored overnight or even only for a few hours can and should be protected by preparing them at a density of 36°Bé. The high-density sirup can, if desired, be diluted to the desired density before use.

3. SUGAR CONCENTRATION AND DILUTION

In the preceding sections, the type of sugar and of sirups used in beverage manufacture and the methods of preparing such sirups have been detailed. In Tables 2-4 and 2-5, the sugar-to-water ratios for the preparation of sirup on a 10-gallon basis and the gallons of sirup of a given Baumé to be obtained with 100 pounds of sugar are tabulated. Such tables avoid hit-and-miss methods of preparation of sirup and simplify calculations.

TABLE 2-4

Sugar: Water Ratio for Ten Gallons of Sirup [a]

°Bé desired	Sugar, lb	Water to be added, gal	Water to be added, oz
20	35	7	48
21	37	7	16
22	40	7	0
23	42	6	112
24	44	6	80
25	46	6	64
26	49	6	48
27	51	6	16
28	53	6	0
29	56	5	96
30	58	5	80
31	60.5	5	48
32	63.5	5	32
33	66	5	0
34	69	4	96
35	72	4	80
36	75	4	48
37	78	4	16
38	82	4	0
39	85	3	96
40	86	3	48

[a] After S. Blumenthal, *Food Manufacturing*, New York, Chemical Publishing, 1942.

TABLE 2-5

Gallons of Sirup Obtained with One-Hundred Pounds of Sugar

°Bé desired	Water to be added		Sirup volume	
	gal	oz	gal	oz
20	20	79	28	13
21	19	8	26	72
22	17	82	25	13
23	16	44	23	102
24	15	20	22	83
25	14	8	21	69
26	13	3	20	70
27	12	7	19	74
28	11	20	18	87
29	10	46	17	107
30	9	74	17	10
31	8	111	16	47
32	8	23	15	84
33	7	68	15	8
34	6	119	14	61
35	6	49	13	118
36	5	106	13	36
37	5	41	12	114
38	4	113	12	52
39	4	56	11	123
40	3	119	11	67

The beverage manufacturer is often confronted with the necessity of determining how much sugar is in his sirup. This can be done by chemical analysis and more simply by specific-gravity measurements, such as using a hydrometer and by reference to tables. Table 2-6 gives the relationship between the Brix (sugar percentage), specific gravity, Baumé hydrometer reading, and the amount of sucrose in 1 gallon of sirup at 60°F for the concentrations most commonly used by soft-drink manufacturers.

At times the beverage manufacturer may have a sirup of higher Baumé on hand, or it may be his custom to prepare sirups of higher Baumé to preserve them, and he may desire to prepare one of lower specific gravity. This can be done by calculation using Table 2-6. By use of Tables 2-7 and 2-8, if he has sirups of 33° and 36° Baumé, the volumes of water to be added for dilution may be obtained directly from the tables.

TABLE 2-6

Weight of Sugar in One Gallon of Sirup at 60°F

°Brix Sucrose % by wt.	Specific gravity	°Bé (modulus 145)	Sucrose, lb	Refractive index 20°C
27	1.114	15	2.500	1.3758
29	1.122	16	2.625	1.3793
31	1.131	17	2.875	1.3829
33	1.140	18	3.000	1.3865
35	1.149	19	3.250	1.3902
36	1.158	20	3.500	1.3920
38	1.167	21	3.625	1.3958
40	1.177	22	3.785	1.3997
42	1.187	23	4.000	1.4036
44	1.196	24	4.250	1.4076
46	1.206	25	4.50	1.4117
48	1.216	26	4.75	1.4158
50	1.226	27	5.00	1.4200
51.5	1.237	28	5.25	1.4232
53.5	1.247	29	5.50	1.4275
55	1.259	30	5.625	1.4307
57	1.270	31	6.000	1.4351
59	1.281	32	6.250	1.4396
61	1.292	33	6.500	1.4441
63	1.304	34	6.750	1.4486
65	1.316	35	7.000	1.4532
67	1.328	36	7.250	1.4579
69	1.340	37	7.500	1.4627
71	1.352	38	7.875	1.4676
73	1.365	39	8.125	1.4725
75	1.379	40	8.500	1.4774

TABLE 2-7

Dilution of 33° Bé Sirup [a]
(1-gallon basis)

°Bé desired	Water to be added, oz
20	110
21	97
22	85
23	74
24	64
25	55
26	46
27	38
28	30
29	23
30	17
31	11
32	6
33	0

[a] Total volume of sirup produced is the sum of 1 gal 33° Bé sirup plus the fluid ounces of water added.

TABLE 2-8

Dilution of 36° Baumé Sirup [a]
(1-gallon basis)

°Bé desired	Water to be added, oz
20	140
21	126
22	112
23	100
24	88
25	78
26	68
27	59
28	51
29	43

TABLE 2-8 (*Continued*)

30	36
31	29
32	22
33	16
34	11
35	5
36	0

[a] Total volume of sirup produced is the sum of 1 gal 36°Bé sirup plus the ounces of water added. Thus for the 20°Bé sirup, the volume would be 1 gal plus 140 oz, yielding 2 gal 12 oz.

4. BEVERAGE-SUGAR CONCENTRATION

The concentration of sugar in a soft drink is governed by the type of the beverage. In Table 2-9, the sugar concentrations usually employed in the more common carbonated beverages are tabulated.

TABLE 2-9

Carbonated-Beverage Sugar Concentrations

Flavor	Sugar, %
Birch beer	10
Celery	9–10
Cherry	11.5–12
Cream soda	11.5–13
Ginger ales	
Golden	9.5–10
Pale dry	8
Grape	13.5–14
Grapefruit	12
Kola	10–10.5
Lemon	11
Lemon and lime	11
Lime	9
Orange	13.5–14
Pineapple	11–12
Raspberry	11–12
Root beer	11–13
Sarsaparilla	9.5–10
Strawberry	12

It can be seen that most of the fruit flavors have relatively high sugar concentrations of the order of 11 to 14 per cent whereas the specialty flavors, like the ginger ales and the kolas, have lower concentrations of the order of 9 to 10 per cent. A pale dry ginger ale has the lowest sugar concentration among the sugar-sweetened beverages.

Sparkling water, club soda, and analogous carbonated beverages are not sweetened.

LITERATURE CITED

1. J. H. Buchanan in *Bottlers' and Glass Packers' Handbook*, p. 51, New York, Food Engineering, McGraw-Hill.
2. Am. Bottlers Carbonated Beverages, "Bottlers' Tentative Sugar Standard," Washington, July 1953.
3. C. Gortatowsky, "Sugar and Other Carbohydrates in Carbonated Beverages," in *Use of Sugars and Other Carbohydrates in the Food Industry (Advances in Chemistry* No. 12), Washington, Am. Chem. Soc., 1955.
4. Am. Bottlers Carbonated Beverages, "Bottlers' Tentative Liquid Sugar Standard," Washington, Jan. 1956.
5. W. B. Newkirk, *Ind. Eng. Chem.*, **28**, 960 (1936); **31**, 18 (1939).
6. *Natl. Bottlers' Gaz.*, **76**, No. 901, 22 (March 1957).

SELECTED REFERENCES

Am. Bottlers Carbonated Beverages, *Plant Operation Manual*, Washington, 1940.

M. B. Jacobs, *Chemical Analysis of Foods and Food Products*, 2nd Ed., New York, Van Nostrand, 1951.

M. B. Jacobs, ed., *Chemistry and Technology of Food and Food Products*, 2nd Ed., New York, Interscience Publishers, 1951.

R. Kühles, *Handbuch der Mineralwasser-Industrie*, Lübeck, Antäus-Verlag, 1947.

Chapter 3

ARTIFICIAL SWEETENERS AND STEVIOSIDE

A. ARTIFICIAL SWEETENERS

The topic of artificial sweeteners is one of endless interest to the beverage manufacturer and the beverage chemist. In 1957 the most important artificial sweeteners were saccharin and cyclamate. While the use of "dietetic" foods (low-calorie foods) has increased greatly since 1940 and both saccharin and cyclamate have been used in dietetic candies, jellies, canned fruits, and similar products, the greatest utilization of sweetening agents, particularly of cyclamate, has been in carbonated beverages. Ever since saccharin was discovered by Remsen and Fahlberg in 1879, the search for such substances has not ceased.[1]

Since, however, no adequate theory has been developed for the relation of sweetness to structure, much of the developments in this field has been purely accidental. This does not mean that we do not know a great deal about structure and sweetness, for much work has been done. Nevertheless, the accidental aspect must be stressed.

1. SACCHARIN

Indeed this has been true of saccharin. Fahlberg noted, after a day's work at his laboratory, that his bread and other food tasted sweet while eating his supper. He went back to his laboratory and checked every test tube, flask, and beaker used in this work until he traced the vessel in which his sweet compound was contained.

$$\text{Saccharin}$$

The principal synthetic sweetening agent for over 50 years has been saccharin and its salts, principally ammonium and sodium salts. Actually the term saccharin is used rather loosely to cover all of these compounds.

Saccharin, itself, is 2,3-dihydro-3-oxobenzisosulfonazole; it is also known as benzoylsulfonimide, o-sulfobenzoic imide, and by a host of trade names, most of which, however, have little use. It is a white crystalline powder which has no odor but a remarkably sweet taste, 300 to 550 times as sweet as sucrose. This rather wide variation in apparent sweetening power is attributable to the wide variation in the sensory response of individual persons. Just as individual persons have wide variations in visual acuity and in hearing, so there is a wide variation in individual taste evaluation. The great sweetening power of saccharin is evident from the fact that many can taste as sweet 1 part of this compound in 70,000 parts of water.

One type of saccharin is known as "550" indicating how many times sweeter than sugar it really is. Saccharin melts with some decomposition at 228°C. It sublimes on further heating. It is not very soluble in cold water, for about 1 gram needs 300 milliliters of cold water to dissolve. One gram dissolves in 25 milliliters of boiling water, in 30 milliliters of alcohol, and in 50 milliliters of glycerol. It is, however, readily soluble in alkaline solutions and in alkali carbonate solutions.

For this reason, saccharin is also sold as its sodium salt, $C_7H_4O_3NSNa \cdot 2H_2O$, under the name of soluble saccharin. This is an odorless, colorless, crystalline powder which is very soluble in water, 83 grams dissolving in 100 milliliters; but only 2 grams dissolve in 100 milliliters of alcohol. This powder effloresces in air. It has about the same sweetening power as saccharin itself. Saccharin is also sold as the ammonium salt and as a solution of either the sodium or the ammonium salt. Finally it is sold also as tablets with added binders and fillers.

Saccharin has been synthesized by a number of methods one of which is from toluene by formation of the o-sulfonic acid. This is converted in turn to the sulfonyl chloride, the amide, and to the o-benzoic acid by potassium permanganate solution. After acidification, water is eliminated and saccharin is formed.

While the alkaline and alkaline-earth salts of saccharin are sweet, the salts formed with heavy metals are bitter. It has been shown that the sweet taste of saccharin disappears when the sulfimide ring is opened and by substituting the hydrogen atom of the imino group with other radicals. Substitution in the aryl nucleus also often results in loss of sweet taste.

Saccharin, as mentioned, has been used for many years for the sweetening of food products, particularly for nonalcoholic beverages like soda pop and in confectionery. It has also been used as a sugar substitute in diabetic diets and for disguising the taste of bitter medicines.

Saccharin has no food value. The Food and Drug Administration has ruled that because saccharin has no food value, its use in food constitutes an adulteration under the Food, Drug, and Cosmetic Act unless the product containing it is clearly labeled to show that it contains a non-nutritive synthetic sweetening agent and is to be used only by those who must restrict their intake of ordinary sweets.

A statement of the ingredients the carbonated beverage contains, followed by some statement concerning the sweetening agent, should be made. An example of a description for such a beverage is the following: "Carbonated water, citric acid, ginger-root extract, essential oils, caramel color, and — per cent of saccharin, a non-nutritive artificial sweetener which should be used only by persons who must restrict their intake of ordinary sweets."

If more than one artificial sweetening agent is used, the label on the bottle or can of carbonated beverage must declare the type of artificial sweetener used, the amounts of each, and pertinent information concerning the other ingredients. For example, "Carbonated water, citric acid, essential oils, artificial color, $1/4$ of 1 per cent cyclamate sodium, $1/100$ of 1 per cent saccharin, non-nutritive artificial sweetening agents which should be used only by persons who must restrict their intake of ordinary sweets."

The Food and Drug Administration considers saccharin in tablet form to be a drug and, therefore, subject to drug labeling requirements in which case stress should be placed on its lack of food value and its displacement of sugar which has food value.

a. **U.S.P. Requirements for Saccharin**

Saccharin is described in U.S.P. XV [2] as follows: Saccharin, dried at 105°C for 2 hours contains not less than 98 per cent of $C_7H_5NO_3S$.

Description—Saccharin occurs as white crystals or as a white, crystalline powder. It is odorless or has a faint, aromatic odor. Saccharin in dilute solution is about 500 times as sweet as sucrose. Its solutions are acid to litmus.

Solubility—One gram of saccharin dissolves in 290 milliliters of water and in 31 milliliters of alcohol. One gram dissolves in about 25 milliliters of boiling water. It is slightly soluble in chloroform and in ether, and is readily dissolved by dilute solutions of ammonia, by solutions of alkali hydroxides, and by solutions of alkali carbonates with the evolution of carbon dioxide.

Identification

A: Dissolve about 100 milligrams of saccharin in 5 milliliters of a solution of sodium hydroxide (1 in 20), evaporate the solution to dryness, and gently fuse the residue over a small flame until it no longer evolves ammonia. Allow the residue to cool, dissolve it in 20 milliliters of water, neutralize the solution with diluted hydrochloric acid, and filter: the addition of a drop of 1 N ferric chloride test solution, prepared by dissolving 9 grams of ferric chloride in sufficient water to make 100 milliliters, to the filtrate produces a violet color.

B: Mix 20 milligrams of saccharin with 40 milligrams of resorcinol, add 10 drops of sulfuric acid, and heat the mixture in a suitable liquid bath at 200°C for 3 minutes. Allow it to cool, and add 10 milliliters of water and an excess of 1 N sodium hydroxide solution: A fluorescent green liquid results.

Melting range—Saccharin melts between 226 and 230°C.

Loss on drying—Dry saccharin at 105°C for 2 hours: It loses not more than 1 per cent of its weight.

Heavy metals—Dissolve 2 grams of saccharin in 4 milliliters 10 per cent ammonium hydroxide solution and add water to make 40 milliliters. Add 1 drop of phenolphthalein indicator solution, prepared by dissolving 1 gram of phenolphthalein in 100 milliliters of alcohol, followed by hydrochloric acid until the pink color is just discharged, then add 2 milliliters of 1 N hydrochloric acid, and immediately add water to make 50 milliliters. Mix well and rub the inner wall of the vessel with a glass rod until crystallization begins. Allow the solution to stand for 1 hour, then filter through a dry filter, rejecting the first 10 milliliters of the filtrate: The heavy metals limit for saccharin, determined in 25 milliliters of the subsequent filtrate is 20 parts per million.

Residue on ignition—Saccharin yields not more than 0.2 per cent of residue on ignition.

Readily carbonizable substances—Dissolve 200 milligrams of saccharin in 5 milliliters of sulfuric acid and keep at a temperature of 48° to 50°C for 10 minutes: The solution has not more color than matching fluid A, which consists of U.S.P. color solutions: 0.1 part cobaltous chloride color solution, 0.4 part ferric chloride color solution, 0.1 part cupric chloride color solution and 4.4 parts of water.

Benzoic and salicylic acids—To 10 milliliters of a hot, saturated solution of saccharin, add 1 N ferric chloride solution dropwise: No precipitate or violet color appears in the liquid.

Assay—Accurately weigh about 500 milligrams of saccharin, previously dried at 105°C for 2 hours, dissolve it in 75 milliliters of hot water, cool quickly, add 3 drops of phenolphthalein indicator solution, and titrate with 0.1 N sodium hydroxide solution. Each milliliter of 0.1 N sodium hydroxide solution is equivalent to 18.32 milligrams of saccharin.

b. **U.S.P. Requirements for Soluble Saccharin**

Soluble saccharin or saccharin sodium is described in U.S.P. XV [2] as follows: Saccharin sodium, dried at 120°C for 4 hours, contains not less than 98 per cent of anhydrous saccharin sodium ($C_7H_4NO_3SNa$).

Description—Saccharin sodium occurs as white crystals or as a white, crystalline powder. It is odorless or has a faint, aromatic odor, and has an intensely sweet taste even in dilute solutions. Saccharin sodium in dilute solution is about 500 times as sweet as sucrose. When in powdered form it usually contains about one-third the theoretical amount of water of hydration as a result of efflorescence.

Solubility—One gram of saccharin sodium dissolves in 1.5 milliliters of water and in about 50 milliliters of alcohol.

Identification
A: Saccharin sodium responds to the identification tests for saccharin.
B: The residue obtained by igniting saccharin sodium responds to the tests for sodium.
C: To 10 milliliters of a solution of saccharin sodium (1 in 10) add 1 milliliter of hydrochloric acid: A crystalline precipitate of saccharin is formed which, when well washed with cold water and dried at 105°C for 2 hours, melts between 226° and 230°C.

Alkalinity—A solution of saccharin sodium (1 in 10) is neutral or alkaline to litmus, but produces no red color with phenolphthalein indicator solution.

Water—Dry saccharin sodium at 120°C for 4 hours: It loses not more than 15 per cent of its weight.

Benzoate and salicylate—To 10 milliliters of a solution of saccharin sodium (1 in 20) previously made acid with 5 drops of acetic acid, add 3 drops of 1 N ferric chloride solution: No precipitate or violet color appears in the liquid.

Heavy metals—Dissolve 2 grams of saccharin sodium in 48 milliliters of water, add 2 milliliters of 1 N hydrochloric acid, mix well, and rub the inner wall of the vessel with a glass rod until crystallization begins. Allow the solution to stand for 1 hour, then filter through a dry filter, rejecting the first 10 milliliters of filtrate: The heavy metals limit for saccharin sodium, determined in 25 milliliters of the subsequent filtrate is 20 parts per million.

Readily carbonizable substances—Saccharin sodium meets the requirements of the test as mentioned under saccharin.

Assay—Weigh accurately about 500 milligrams of saccharin sodium, previously dried at 120°C for 4 hours, and transfer it completely to a separatory funnel with the aid of 10 milliliters of water. Add 2 milliliters of diluted hydrochloric acid and extract the precipitated saccharin first with 20 milliliters and then with four 20-milliliter portions of a solvent composed of 9 volumes of chloroform and 1 volume of alcohol, filtering each extract through a small filter paper moistened with the solvent. Evaporate the combined filtrates to dryness on a steam bath with the aid of a current of air, then dissolve the residue in 75 milliliters of hot water, cool, add 3 drops of phenolphthalein indicator solution, and titrate with 0.1 N sodium hydroxide solution. Each milliliter of 0.1 N sodium hydroxide solution is equivalent to 20.52 milligrams of saccharin sodium.

c. **Formulation**

It may be pointed out here that a beverage containing saccharin, cyclamate, or a combination of both, is not considered a soft drink in the strict sense of the words. Such artificially sweetened beverages are not labeled, for instance, as "Ginger Ale" but as "Ginger Flavored Beverage," nor as "Orange Soda" but as "Orange Flavored Beverage," etc.

It has already been stressed that artificial sweeteners and, in this particular instance, saccharin is permitted only in beverages to be used by persons who must restrict their intake of ordinary sugars and sweets. It must also be stressed that in the United States, saccharin and other artificial sweetening agents are not allowed as partial replacements for sugar by the Food and Drug Administration or by the various State control agencies. In some foreign countries, saccharin can be used as the sole sweetening agent in a beverage or as a partial replacement for sugar. In the first case, as is discussed in greater detail in a subsequent section of this chapter, bodying agents may be used to give the beverage the "body" the beverage lacks if sugar is not used. The replacement values of saccharin for sucrose in terms of grams, or grains, or ounces is given in Table 3-1 and the sugar equivalent of mixtures of saccharin and sugar is given in Table 3-2.

TABLE 3-1

Replacement of Sugar by Saccharin

Sugar lb	Saccharin "550" equivalents		
	g	gr	oz
1	0.91	14	0.03
5	4.50	70	0.16
10	9.10	140	0.32
25	22.50	350	0.80
50	45.00	700	1.60
100	91.00	1400	3.20

TABLE 3-2

Sugar and Saccharin Mixtures [a]

Saccharin + Sugar		Sugar equivalent
oz	lb	lb
0.25	8.75	17.5
0.5	17.5	35
1.0	35	70
2.0	70	140

[a] Saccharin and/or other artificial sweeteners or combinations of artificial sweeteners are not permitted as partial replacement of sugar by the Food and Drug Administration and by State control agencies.

2. CYCLAMATE

Sodium cyclohexylsulfamate was also discovered accidentally for it was in a study of the antipyretic properties of the sodium salt of N-phenylsulfamic acid, $C_6H_5NHSO_3Na$, and related compounds that it was noted that the salts of N-cyclohexylsulfamic acid were remarkably sweet. This, in turn, led to an investigation of the relation of structure to sweetness by Audreith and Sveda.[3]

$$\begin{array}{c} CH_2 \\ H_2C \quad CH-N-SO_2-ONa \\ | \quad | \quad | \\ H_2C \quad CH_2 \quad H \\ CH_2 \end{array}$$ Sodium Cyclohexylsulfamate

a. Sodium Cyclamate

The work of these investigators indicated that sodium cyclohexylsulfamate, which is called cyclamate and sulfamate for short by some workers, and is also known under the name of sodium cyclamate and the trademarks Sucaryl sodium (Abbott Laboratories)[4] and Cylan sodium cyclamate (Du Pont),[5] is a very sweet substance. Originally it was thought to be about 100 times as sweet as sucrose, but greater experience with this product shows that it is only about 30 times as sweet as sucrose.

Sodium cyclohexylsulfamate is a white, free-flowing nonhygroscopic, noncorrosive, and practically odorless solid that has no nutritive value. It is compatible with virtually all ingredients used in making foods and beverages, it has no effect on the shelf life of beverages or foods, and retains its sweetness indefinitely. It is stable in the pH range of 2 to 10 and is not affected by temperatures up to 500°F.

The specifications of Du Pont are for a product containing not less than 98 per cent of sodium cyclohexylsulfamate; less than 20 ppm of heavy metals; not more than 1 per cent loss on drying; and a pH of a 10 per cent aqueous solution in the range of 5.5 to 7.5.

Sodium cyclamate is very soluble in water, forming clear colorless solutions. The solubility in weight percentage is at 50°F (10°C) 15 per cent, 77°F (25°C) 18 per cent, 104°F (40°C) 28 per cent, 131°F (55°C) 40 per cent, and 158°F (70°C) 44 per cent.

(1) N.N.R. Requirements [6]

Physical properties—Sodium cyclamate is a white, crystalline, practically odorless powder with a very sweet taste. It is freely soluble in water and practically insoluble in alcohol, benzene, chloroform, and ether. The pH of a 10 per cent solution of sodium cyclamate is between 5.5 and 7.5.

Identity tests—To about 15 milligrams of sodium cyclamate in 10 milliliters of water, add 1 milliliter of hydrochloric acid, 20 milligrams of sodium nitrite, and 0.1 gram of barium chloride: A white precipitate forms.

To 1 milliliter of a 10 per cent solution of sodium cyclamate, add 2 milliliters of 0.1 N silver nitrate solution and shake: A white, crystalline precipitate forms.

Purity tests—Determine the heavy metals content on 1 gram of sodium cyclamate as directed in Chapter 4, section 2.a, subsection heavy metals: The turbidity developed does not exceed that of a control containing 20 micrograms of lead.

Dry about 1 gram of sodium cyclamate, accurately weighed, for 1 hour at 105°C: The loss in weight is not more than 1.0 per cent.

Assay—Nitrogen—Transfer about 0.1 gram of sodium cyclamate, accurately weighed, to a semimicrokjeldahl flask and proceed with the method as detailed for calcium cyclamate. Each milliliter of 0.02 N acid is equivalent to 0.0002802 gram of nitrogen and

0.004024 gram of sodium cyclamate. The amount of nitrogen present is not less than 7.1 per cent, equivalent to not less than 98.0 nor more than 102.0 per cent of sodium cyclamate.

b. **Calcium Cyclamate**

Another salt of cyclohexylsulfamic acid that is very sweet is the calcium salt $(C_6H_{11} \cdot NHSO_3)_2Ca \cdot 2H_2O$, that is, calcium cyclohexylsulfamate dihydrate. It is a white, crystalline, nonhydroscopic, free-flowing solid that is about 30 to 40 times as sweet as sucrose. It has no nutritive value. It is sold under trademarks, such as Sucaryl calcium (Abbott Laboratories) [4] and Cylan calcium cyclamate (Du Pont).[5]

The specifications of Du Pont are for a product containing not less than 98.4 per cent of calcium cyclamate dihydrate, based on calcium analysis; calcium not less than 9.1 per cent; heavy metals less than 5 ppm; benzene-extractable matter less than 10 ppm; pH of aqueous solutions in the range of 5.0 to 7.5; and crystal size, less than 10 per cent through a 100-mesh sieve.

Calcium cyclamate is very soluble in water, forming clear, colorless solutions. The solubility in weight percentage is at 32°F 17 per cent, 77°F 20 per cent, 122°F 40 per cent, 167°F 50 per cent, and 212°F 55 per cent. It is soluble in aqueous alcohol at room temperature: 25 grams per 100 milliliters of 25 per cent alcohol, decreasing to 2.4 grams in 100 milliliters of 90 per cent ethyl alcohol. It is also soluble in aqueous solutions of sorbitol.

Calcium cyclamate reacts with tartaric acid and tartrates, oxalates, and some phosphates to form insoluble calcium salts. Consequently, it is preferable to use the sodium salt of cyclohexylsulfamic acid in those formulations where a precipitate might form.

(1) N.N.R. Requirements [6]

Physical Properties—Calcium cyclamate is a white, crystalline, practically odorless powder with a very sweet taste. It is freely soluble in water and practically insoluble in alcohol, benzene, chloroform, and ether. The pH of a 10 per cent solution is between 5.5 and 7.5.

Identity Tests—To 1 milliliter of a 15 per cent solution of calcium cyclamate, add 2 milliliters of 0.1 N silver nitrate solution and mix: A white crystalline precipitate forms.

To 2 milliliters of a 15 per cent solution of calcium cyclamate, add 1 milliliter of 10 per cent sodium nitrite solution, shake, and cautiously add hydrochloric acid dropwise: A white precipitate forms.

Purity Tests—Determine the heavy-metals content on 1 gram of calcium cyclamate as directed in Chapter 4, section 2.a, subsection heavy metals: The turbidity developed does not exceed that of a control containing 20 micrograms of lead.

Dry about 1 gram of calcium cyclamate, accurately weighed, for 4 hours at 130°C: The loss in weight is not more than 9.5 per cent.

Assay—(Nitrogen) Transfer to a semimicrokjeldahl flask about 0.1 gram of calcium cyclamate, accurately weighed. Digest it with 7 milliliters of sulfuric acid, 2.7 grams of potassium sulfate, and 0.3 gram of mercuric oxide for 4 hours. Cool the contents of the flask and add 15 milliliters of water. Make the solution alkaline with 40 per cent sodium hydroxide solution and add 10 milliliters of 40 per cent sodium thiosulfate solution. Distill the ammonia formed into 20 milliliters of 4 per cent boric acid. Using a mixed indicator (1:5 methyl red indicator solution (0.1 gram in 100 milliliters of ethyl alcohol) and bromocresol green indicator solution (50 milligrams in 100 milliliters of ethyl alcohol), titrate the ammonia liberated with 0.05 N sulfuric acid. Each milliliter of the 0.05 N acid is equivalent to 0.0007004 gram of nitrogen and 0.009913 gram of anhydrous calcium cyclamate. The amount of nitrogen present is not less than 6.93 nor more than 7.21 per cent, equivalent to not less than 98.0 nor more than 102.0 per cent of anhydrous calcium cyclamate.

(Calcium) Transfer to a 400-milliliter beaker about 0.5 gram of calcium cyclamate, previously dried at 130°C for 4 hours and accurately weighed. Add 10 milliliters of diluted hydrochloric acid and 15 milliliters of 0.5 N ammonium oxalate solution, and heat the mixture just to boiling. Add 25 milliliters of 10 per cent ammonium hydroxide solution, continue boiling for 5 minutes, allow the beaker to stand for 1 hour, then filter the mixture through a sintered-glass crucible. Wash the residue with about 200 milliliters of water. Transfer the crucible containing the precipitate back to the original beaker, add enough water to cover the crucible and 7 milliliters of sulfuric acid, heat the contents of the beaker to 90°C and then titrate with 0.1 N potassium permanganate solution. If the temperature of the solution falls below 70°C, reheat it before continuing the titration. Each milliliter of 0.1 N potassium permanganate solution is equivalent to 0.002004 gram of calcium and 0.01983 gram of anhydrous calcium cyclamate. The amount of calcium present is not less than 9.91 nor more than 10.3 per cent, equivalent to not less than 98.0 nor more than 102.0 per cent of anhydrous calcium cyclamate.

c. **Other Sulfamates**

Ammonium cyclohexylsulfamate is so remarkably sweet, that a dilution of 1:5000 can be detected. Sodium o-methylcyclohexylsulfamate, $o\text{-}CH_3C_6H_{10}NHSO_3Na$, is also very sweet, for it can be detected in a ratio of 1:2500.

d. **Toxicity**

In a series of tests, conducted by the Food and Drug Administration,[7] on the relative toxicity of saccharin, sodium cyclohexylsulfamate, 2-amino-4-nitro-1-n-propoxybenzene (P-4000), and dulcin, it was found that dulcin produces liver tumors and interferes with the production of red cells in animals and is, therefore, excluded on the basis of harmful effects on long-term low-level feeding. P-4000 produced kidney damage and an indication of interference with thyroid function, thus it is excluded on the basis of a low margin of safety. However, both saccharin and sodium cyclohexylsulfamate produced no apparent effects at high levels of feeding and, consequently, are considered safe as food additives.

One other factor to consider with all of these compounds is that none of them has any food value so that their use in food constitutes an adulteration unless the product containing the permitted sweetening agent is clearly labeled to show that it contains a nonnutritive synthetic sweetening agent. It is also wise to restrict their use principally to those whose intake of ordinary sweets must be limited. In some communities, the use of artificial sweeteners is prohibited in certain products.

e. **Formulation**

In the preceding chapter the preparation of acidified and flavored sirups was discussed. Comparable sweet concentrates of artificial sweeteners must also be prepared. A number of factors are involved in the manufacture of such concentrates. First, compatibilities must be considered. For example, if tartaric acid is to be used as the acidulant, then it is preferable to use sodium rather than calcium cyclamate. Second, the apparent acidity, with respect to taste, of citric acid is higher with cyclamates in a formulation than with sugar. It has been suggested that 10 to 20 per cent less acid is used with cyclamates. Third, it is well to recall that sugars give not only sweetness to a carbonated beverage but also body. Therefore, in artificially sweetened beverages, the use of bodying agents should be considered (see Chapter 8, section 5). Fourth, if sorbitol is used as a bodying agent, it must be remembered that sorbitol has sweetening power also and, therefore, the amount of cyclamate used in the formulation must be reduced. Fifth, if cyclamate concentrates are used in making cloudy beverages, such as citrus beverages, the specific gravity of the emulsion concentrate must be adjusted at lower levels than with sugar solutions if sorbitol is used and at still lower levels if another or no bodying agent is used. This is discussed in detail in Chapter 8, section 4.e.

With these factors in mind, the representative cyclamate concentrate formulations in Table 3-3, can be used and modified to suit the market served by the beverage manufacturer.

Soft drinks containing few calories are used by people who wish to or who must restrict their calorie intake and who, at the same time, wish to have a palatable and refreshing beverage. In order to comply with the regulations of the Food and Drug Administration and with the regulations of other regulatory agencies, it is necessary for "dietetic" artificially sweetened soft drinks to be labeled properly when offered for sale. In general, this means that the label must indicate that the beverage is designed for dietetic purposes and must give the name of the artificial non-nutritive sweetener. The chemical analysis of the product showing the percentage of carbohydrates, proteins, and fat, and the number of calories per average serving must be stated. Since there is some variation in state, county, and municipal regulations, it is incumbent on the beverage manufacturer to acquaint himself with such rules and regulations and give the information required by them on his label. Examples of proper labeling were given in connection with the discussion of this topic for saccharin. See this chapter, section A.1.c.

3. DULCIN

It would be unrealistic to fail to mention other artificial sweetening agents for though they are not approved for use in the United States and are not looked on with favor in other countries, they may

TABLE 3-3
Cyclamate Concentrate Formulations[a]

FLAVOR	Pale dry ginger ale		Kola		Orange	
Bottlers Concentrate	Amount	% by wt.	Amount	% by wt.	Amount	% by wt.
Ca cyclamate	14 lb	1.66	20 lb	2.38	25 lb	3.00
Citric acid, cryst.	10 lb	1.19	7 lb	0.83	12 lb	1.44
Flavor Extract[b]	200 fl oz	1.55	25 pt	3.10	200 fl oz	1.56
Water	96⅔ gal	95.60	94⅔ gal	93.69	95½ gal	94.00
Total	100 gal	100.00	100 gal	100.00	100 gal	100.00
Throw to 12-oz bottle:		2 fl oz		2 fl oz		2 fl oz
Carbonate at:		4.5 volumes		4 volumes		2.5 volumes
Beverage Composition		% by wt.		% by wt.		% by wt.
Ca cyclamate		0.28		0.40		0.50
Citric acid		0.20		0.14		0.24
Flavor		0.26		0.50		0.26
Water		99.26		98.96		99.00
Bodying Agents						
If desired, use:						
CMC 70 S High[c]	19 oz		27 oz		34 oz	
or use:						
Sorbitol (70% solution)	7 gal		8½ gal		10 gal	
Reduce "Cylan" to	12 lb		17½ lb		22 lb	
Preservative						
If necessary, use						
Benzoic acid	12 oz		12 oz		12 oz	

[a] After E. I. Du Pont deNemours, Technical Data Sheet No. 1, "Cylan" Calcium Cyclamate (1957).
[b] Based on usual strengths of prepared extracts, intended to be used as follows: Ginger ale, 2 fl oz per gal of sirup; kola, 4 fl oz per gal of sirup (also assuming the extract contains no acid; if it contains acid, the citric acid must be reduced or omitted entirely); orange, 2 fl oz per gal of sirup. If other strengths of extracts are used, the amounts must be increased or decreased accordingly in the listed formulas. Some prepared bottlers' extracts contain the necessary color. Otherwise, color must be added to the sirup as required.

not be found objectionable in some countries. One need only mention dulcin which was used for over 50 years as an artificial sweetening agent before reinvestigation by the Food and Drug Administration, as mentioned, led to the conclusion that its toxicity was of such an order that it would be in the best interest of public health to bar its use in foods. Jacobs [8] has discussed artificial sweeteners in great detail.

Dulcin, also known as p-phenetylurea, p-phenetyl carbamide, Sucrol, and by other names, $NH_2CONHC_6H_4OC_2H_5$, is a solid crystallizing in white lustrous crystals. It melts at 173° to 174°C. It is only slightly soluble in water; only 1 gram dissolves in 800 milliliters of cold and 50 milliliters of boiling water. It is much more soluble in alcohol, 1 gram dissolving in 25 milliliters of alcohol. This artificial sweetening agent has about 200 times the sweetening power of sucrose. One of the reasons that dulcin remained a competitor of saccharin for many years was that its sweet taste was more agreeable than that of saccharin. A derivative of dulcin, beta-4-ethoxyphenyl-beta-methylurea was shown to be sweeter than dulcin and more soluble in water. Neither dulcin nor its derivatives have any food value.

4. ALKOXYAMINONITROBENZENES

The 1-alkoxy-2-amino-4-nitrobenzenes were proposed as artificial sweeteners by Verkade in 1946. Initially they were received with great favor for they were much more powerful as sweetening agents than saccharin, so that much less had to be used and, in addition, they had no bitter aftertaste. The 1-n-propoxy-2-amino-4-nitrobenzene or P-4000 as it was called, because it was considered to be 4000 times as sweet as sucrose seemed to have great promise, but, as mentioned previously, tests of the Food and Drug Administration indicated that it might have too high toxicity. Consequently it has not been approved for food use.

B. NATURAL HIGH-POWER SWEETENING AGENTS

There are a number of natural substances of high sweetening power which may be of interest to the beverage manufacturer. The most important ones are perilla aldehyde antialdoxime and stevioside.

1. PERILLA ALDEHYDE ANTIALDOXIME

Many years ago, Semmler isolated an unsaturated aldehyde from

$$OHC-C\begin{matrix}CH-CH_2\\ \\CH_2-CH_2\end{matrix}CH-C\begin{matrix}CH_2\\ \\CH_3\end{matrix}$$

an essential oil of a plant called *Perilla nankinensis*, very likely more properly, *Perilla frutescens* which is the source of genuine perilla oil and which contains 40 to 55 per cent of perillaldehyde, or perilla aldehyde as it is often termed.

The flavor of the oil and perillaldehyde is not significantly sweet, but the alpha-*anti*-aldoxime, prepared by Furukawa and Tomizawa about 1920, is 2000 times as sweet as sucrose and thus from 4 to 8 times as sweet as saccharin. Since the alpha-*anti*-aldoxime is a derivative of an isolate, it is not really a natural product, but the relative change or processing is not intensive.

It is of interest to point out here that the isomeric beta-*syn*-aldoxime is not sweet at all. This is analogous to the finding of Gilman and Dickey in 1930 that the synthetic compound *syn*-5-benzyl-2-furaldoxime was sweet, about 700 times as sweet as sucrose, whereas the *anti* isomer was only 100 times as sweet as sugar.

Another note of interest is that the Japanese Government permits the use of perilla alpha-*anti*-aldoxime as a replacement for licorice, honey, or maple syrup in the sweetening of tobacco and that most of the perilla oil produced in Japan is converted to this sweetening agent for that purpose.

2. STEVIOSIDE

Recently, Fletcher, of the National Institute of Arthritis and Metabolic Diseases, National Institutes of Health, Public Health Service, United States Department of Health, Education, and Welfare, focused attention again on stevioside, a glycoside that can be extracted from a Paraguayan plant. In 1954, Bell also called attention to this unique sweetening agent.

The plant from which this sweetening agent, which is said to be 300 times as sweet as sucrose, has been extracted is presently known as *Stevia rebaudiana* Bertoni, Family of Compositae, and was originally given the name *Eupatorium rebaudianum*. The natives of Paraguay, where the shrub is grown or found, have used it to sweeten bitter beverages for many years. It is called in the Guarani dialect *ca-á jhëë (kaà hê-ê)* or *ca a yupe*.

It is possible that the sweet character of this plant was noted by Bertoni in 1899 and very likely Rasenack noted it later in 1908, but it was not until the work of Briedel and Lavielle, in 1931, that its remarkably sweet character was fully recognized.

These investigators isolated the glucoside from the leaves by means of aqueous alcoholic extraction. They were able to crystallize the compound, but were unable originally to hydrolyze it with the enzymes of air-dried brewers' yeast or the powder of *Aspergillus niger*, emulsin, or rhamnodiastase. Later they were able to hydrolyze the glycoside enzymically by using the hepato-pancreatic juice or the digestive juice of the vineyard snail, *Helix pomatia*. They were also able to hydrolyze the compound with 5 per cent sulfuric acid at 100°C. This processing gave them two products, one being D-glucose and the other being the aglucon that is the nonsugar part of the molecule.

Subsequently Briedel and Lavielle studied the physiological response to this sweetening agent. They found that this glucoside did not appear to be toxic to guinea pigs, rabbits, or chickens and that it was a neutral substance. Since that time very little was done on elucidating the structure of the compound although, as Wood and his collaborators point out, a number of authors did call attention to the powerful sweetening effect of this substance.

In 1952, the study of this compound was resumed by investigators of the United States Public Health Service as a part of a general investigation of natural products that might prove useful as raw materials.

By more modern methods as much as 7 per cent of crude stevioside was isolated from the dried leaves. This was done by extracting the dried plant with water and deionizing the extract with ion-exchange resins as in modern methods of beet-sugar refining. The resultant solution was concentrated *in vacuo* to give a residue from which the crude stevioside was crystallized with the aid of methyl alcohol. The pure glycoside was obtained by recrystallization from a mixture of methyl alcohol and dioxane.

The structure of the glycoside was studied by two groups

of investigators, one of which studied the glucose fractions and the other the aglucon portion. It was found to have the formula $C_{12}H_{21}O_{11} \cdot C_{19}H_{28}CO \cdot C_6H_{11}O_6(C_{38}H_{60}O_{18})$.

The stevioside molecule contains three glucosepyranose units, two of which are joined together in an unusual linkage and then to the aglucon while the third is esterified to the aglucon moiety. The aglucon obtained by enzymic hydrolysis is known as steviol while that obtained by acid hydrolysis is known as isosteviol. Both appear to be diterpenoid acids with a 2,11-cyclopentanoperhydrophenanthrene skeleton.

Fletcher, in appraising the possibility of the use of stevioside as a natural noncaloric sweetening agent, was not optimistic as to its possibility of competing with such a relatively inexpensive artificial sweetening agent as saccharin. He points out that assuming that the yield of an acre of the experimenters in Paraguay could be duplicated, then the yield per acre would be 53 pounds if the leaves could produce 6 per cent of the glycoside. Since this would be equivalent to about 40 pounds of saccharin at about $90.00 the stevioside would have to show marked flavor properties over saccharin. This might be possible since it is said not to have the bitter after-taste of saccharin.

However, Fletcher stresses the need for caution in using stevioside before really adequate testing of its toxicity and physiological response in human beings. Although there are no reports of harm to the Paraguayan users of this sweetening agent, only after exhaustive and long-term tests of the effect of the ingestion of stevioside have been made can the suitability for food use be evaluated.

LITERATURE CITED

1. M. B. Jacobs, *Am. Perfumer Essent. Oil Rev.* **57**, No. 1, 49 (1951).
2. *The Pharmacopeia of the United States of America* (The United States Pharmacopeia) *Fifteenth Revision,* U.S.P. XV, U.S. Pharmacopeial Convention, Inc., Washington; Easton, Pa., Mack Publishing, Dec. 1955.
3. L. F. Audreith and M. Sveda, *J. Org. Chem.,* **9**, 89 (1944).
4. "Making the Most of the Dietetic Market with 'Sucaryl,'" North Chicago, Abbott Laboratories, 1956.
5. "Cylan" (Technical Data Sheets No. I, III, and VIII), Wilmington, E. I. Du Pont de Nemours & Co., (Inc.), Grasselli Chemicals Dept. 1957.

6. *New and Nonofficial Remedies,* Am. Med. Assoc., Philadelphia, Lippincott, 1952.
7. A. J. Lehman, *Assoc. Food Drug Officials (U. S.) Quarterly Bull.,* **14,** 82 (1950).
8. M. B. Jacobs, Am. Perfumer Essent. Oil Rev., **57,** No. 1, 49 (1951); **57,** No. 2, 129 (1951); **62,** 295 (1953); **65,** No. 2, 56 (1955); and **66,** No. 6, 44 (1955).

SELECTED REFERENCES

"Sucaryl" Sweetened Beverages, Technical Information No. 42, North Chicago, Abbott Laboratories, 1957.

K. M. Beck, *Food Eng.,* **26,** 87 (June 1954).

M. B. Jacobs, *Synthetic Food Adjuncts,* New York, Van Nostrand, 1947.

R. H. Morgan, *Beverage Manufacture,* London, Attwood & Co., 1938.

Chapter 4

ACIDS AND ACIDULATION

The flavor and quality of carbonated nonalcoholic beverages are dependent in some measure on the amount and character of the acid added for the flavor is developed in part and complemented in part by the acidulant. The principal acids used in the manufacture of soda pop are citric, phosphoric, and tartaric acids. Lactic and malic acids are also used but less extensively. On occasion, acetic acid may be used and the literature has some references to the use of other acids, for instance, adipic acid is used in Europe. All of these acids, with the exception of phosphoric acid, are organic acids. All the acids used in beverage manufacture, including phosphoric acid, must be of edible grade.

1. ACIDITY

In evaluating the acidity of any beverage, two aspects must be considered. These are (1) the quantity aspect, that is, the total amount of available hydrogen ion present in a given volume of the beverage as estimated by a determination of the total titratable acidity and (2) the intensity aspect, that is, the apparent hydrogen-ion concentration usually expressed in terms of pH.

Any substance in aqueous solution is either acid or alkaline, depending on whether it yields an excess of hydrogen ions or an excess of hydroxyl ions. The hydrogen-ion concentration may vary over a very wide range, for instance, from 1 gram of ions per liter to less than 1 million-millionth. Expressing these concentrations in ordinary concentration terms results in very unwieldy expressions, such as, in the case of a cherry soda, 1/1000 of a gram of hydrogen ion in one liter of the beverage. Actually it seldom occurs that the value of the hydrogen-ion concentration is an even decimal fraction, such as 1/1000 or 1×10^{-3}. It is much more likely that some random value would be obtained. For this reason it has become common to express acidity in pH terms.

a. pH

The symbol, pH, has been adopted for the logarithm of the reciprocal of the hydrogen-ion concentration. If the hydrogen-ion concentration (represented by [H$^+$]) of a solution is known, the corresponding pH of the solution may be calculated from the formula:

$$p\text{H} = \log\frac{1}{[\text{H}^+]}$$

It is not really necessary to consider the meaning of pH in terms of the theory of solution. The pH numbers need only be accepted as a practical scale of acidity and alkalinity with a pH of 7.00 being the neutral point, that is, the point where the concentrations of hydrogen ion and hydroxyl ion are equal. Solutions in which the hydrogen-ion concentration is greater than the hydroxyl-ion concentration, that is, acid solutions, have lower pH values ranging down to 1.0 or lower. Solutions in which the hydroxyl-ion concentrations is greater than the hydrogen-ion concentrations are alkaline, and the pH is expressed by higher values ranging up to 13 or 14.

In Table 4-1, the pH, the percentage of acid, and the acid employed in some of the more common carbonated beverages are tabulated. It can be seen from this table that these beverages can be placed into two groups characterized by their acid content and pH. Thus the acid-type carbonated beverages comprise the ginger ales, kolas, and fruit-flavored beverages, whereas the nonacid type include the root beers, sarsaparillas, and the sparkling waters.

TABLE 4-1

Acid Concentration of Carbonated Beverages
and Fruit Juices

Flavor	Carbonated beverages			Fruit juices		
	Acid, %	pH	Acid used	Acid, % avg.	pH approx.	Acid used for calcn.
Apple	0.10	3.5	malic, citric	0.5	3.3	malic
Blackberry	0.12	3.0	citric	1.4	3.0	malic
Cherry	0.10	3.0	citric, malic	1.2	3.2	malic
Grape	0.10	3.0	tartaric	1.2	3.0	tartaric
Pineapple	0.12	3.3	citric	0.8	3.2	citric
Raspberry	0.10	3.0	citric	1.3	3.6	malic
Strawberry	0.10	3.0	citric	1.1	3.3	malic
Grapefruit	0.18	3.0	citric	1.5	3.2	citric
Lemon	0.13	2.7	citric	6.0	2.2	citric
Lime	0.14	2.6	citric	7.0	2.0	citric
Orange	0.08	3.5	citric	1.2	3.3	citric
Celery	0.02	5.7	malic, citric			
Cream soda	0.02	5.5	citric			
Ginger ale						
Golden	0.12	2.8	citric			
Pale dry	0.24	2.6	citric			
Kola	0.05	2.3	phosphoric			
Root beer	0.01	5.0	phosphoric			
Sarsaparilla	0.01	4.5	phosphoric			
Tom Collins	0.30	3.2	citric			

The pH of pure water is 7.0, but this is true only of specially purified electrolytic water. For instance, ordinary distilled water has a pH of the order of 5.8 to 6.0 when it is in equilibrium with the air. Usually tap water has a wider range and may even be alkaline, that is, have a pH above 7. This depends on the source, treatment, and the nature and quantity of dissolved salts. Carbonated beverage water will be considered in detail in the next chapter.

b. **Normality**

The quantity of acid is expressed in terms of normality, percentage, or grains per gallon. The last expression used to be the most common, but with increase in the number of persons in the beverage industry who have studied chemistry, the use of normality is also becoming wide spread. Acids prepared and sold on a percentage basis are also common, as for instance, 50 per cent citric acid, 85 per cent phosphoric acid, 75 per cent phosphoric acid, 50 per cent lactic acid, and 80 per cent lactic acid.

A normal solution is one which contains one gram-molecular weight in a liter of solution. In general, the normality of a solution is its concentration in multiples or fractions of the number of equivalent gram-molecular weights of solute it contains per liter.

The normality of the various acids used in the beverage industry can all be adjusted to the same value, but at this normality the pH may be considerably different. Thus, for instance, a 0.02 N solution of phosphoric acid at 25°C in 10 per cent sugar solution has a pH of 2.45 whereas the same concentration of lactic acid in the same concentration of sugar solution has a pH of 2.73. Some of these differences are shown in Table 4-2.

TABLE 4-2

Hydrogen-Ion Concentration of Carbonated Beverage Acids [a]

Acid	pH 0.02 N in 10% sucrose soln. at 25°C	Ionization constant K	Dissociation, % approx.
Phosphoric	2.45	1.1×10^{-2}	70
Tartaric	2.58	1.1×10^{-3}	30
Malic		4.0×10^{-4}	
Citric	2.68	8.0×10^{-4}	28
Lactic	2.73	1.6×10^{-4}	9
Acetic	3.50	1.75×10^{-5}	3

[a] After M. Sivetz, in *Bottlers' and Glass Packers' Handbook*, p. 29, New York, Food Engineering, McGraw-Hill.

2. CITRIC ACID

Citric acid, $COOH \cdot CH_2 C(OH)(COOH) CH_2 COOH \cdot H_2O$, beta-hydroxytricarballylic acid, is a practically odorless, colorless solid, forming either translucent crystals, or white granules, or powder. It was first isolated by Scheele in 1784 from lemon juice. The hydrated acid loses water when it is heated at 70° to 75°C, and has an apparent melting point at 100°C, but the anhydrous acid actually melts at 153°C. The hydrate has a density of 1.542. It is very soluble in water, approximately 133 parts dissolving in 100 parts of cold water, and in ethyl alcohol, approximately 76 grams dissolving in 100 milliliters of absolute alcohol at 15°C. Its calcium salt is only slightly soluble in cold water and is even less soluble in hot water. While citric acid is the fruit acid characteristic of the citrus fruits, it is also a component of many other fruits like cranberry, currant, pineapple, and berry fruits like strawberry and raspberry. Citric acid has a pleasant sour taste and a flavor reminiscent of lemon. It can be prepared synthetically, but most commercial citric acid is made by a fungal fermentation of sucrose and some is obtained by extraction from lemons, limes, and pineapples. Among the molds used are species of *Aspergillus,* particularly *Aspergillus niger, Penicillium,* particularly the monoverticillate penicillia with which the *Citromyces* were formerly classified, and *Mucor.* In its manufacture from fruit juices, the citric acid is precipitated as the calcium salt, which is relatively insoluble as noted before, filtered off, redissolved by the addition of sulfuric acid with precipitation of calcium sulfate, filtered again. The acid is then obtained by crystallization from the concentrated solution.

Citric acid is the principal acid used in the carbonated-beverage industry. The major reason for its wide employment is the fact that it combines well with fruity and light flavors. The 50 per cent aqueous solution is the stock solution most commonly used. Although the stock solutions of citric acid are commonly called 50 per cent, in the majority of instances, 5 pounds of citric acid hydrate is added to sufficient water to make 1 gallon of solution. This solution contains actually 49.25 per cent citric acid. If the bottler purchases the anhydrous acid, he generally uses 4 pounds and 9 ounces in sufficient water to make 1 gallon.

In Table 4-3, the Baumé hydrometer readings, specific gravity, grams per liter, and pounds per gallon corresponding to given percentages of citric acid solutions are tabulated.

a. **United States Pharmacopeia Requirements**

The United States Pharmacopeia XV [1] details the following criteria for citric acid, U. S. P. XV.

Citric acid is anhydrous or contains one molecule of water of hydration. It contains not less than 99.5 per cent of $C_6H_8O_7$, calculated on the anhydrous basis.

Description—Citric acid occurs as colorless, translucent crystals, or as a white, granular to fine, crystalline powder. It is odorless,

TABLE 4-3

Citric Acid, Aqueous Solutions
Specific Gravity at 15°C

$C_6H_8O_7 \cdot H_2O$, %	Bé	Sp. Gr.	Grams/liter	Pounds/gallon
2	1.1	1.0074	20.15	0.1681
4	2.1	1.0149	40.60	0.3388
6	3.2	1.0227	61.36	0.5121
8	4.3	1.0309	82.47	0.6883
10	5.5	1.0392	103.9	0.8673
12	6.5	1.0470	125.6	1.049
14	7.6	1.0549	147.7	1.232
16	8.6	1.0632	170.1	1.420
18	9.7	1.0718	192.9	1.610
20	10.8	1.0805	216.1	1.803
22	11.8	1.0889	239.6	1.999
24	12.8	1.0972	263.3	2.198
26	13.9	1.1060	287.6	2.400
28	15.0	1.1152	312.3	2.606
30	16.0	1.1244	337.3	2.815
32	17.1	1.1332	362.6	3.026
34	18.1	1.1422	388.3	3.241
36	19.1	1.1515	414.5	3.460
38	20.1	1.1612	441.3	3.682
40	21.2	1.1709	468.4	3.909
42	22.3	1.1814	496.2	4.141
44	23.1	1.1899	523.6	4.369
46	24.2	1.1998	551.9	4.606
48	25.2	1.2103	580.9	4.848
50	26.2	1.2204	610.2	5.092
52	27.2	1.2307	640.0	5.341
54	28.2	1.2410	670.1	5.593
56	29.1	1.2514	700.8	5.848
58	30.2	1.2627	732.4	6.112
60	31.2	1.2738	764.3	6.378
62	32.2	1.2849	796.6	6.648
64	33.1	1.2960	829.4	6.922
66	34.1	1.3071	862.7	7.199

has a strongly acid taste, and the hydrous form is efflorescent in dry air.

Solubility—One gram of citric acid dissolves in 0.5 milliliter of water, in 2 milliliters of alcohol, and in about 30 milliliters of ether.

Identification—A solution of citric acid responds to the tests for citrate. To a solution of 5 milliliters of a citrate (1 in 10) add 1 milliliter of 1 N calcium chloride solution and 3 drops of bromothymol blue indicator solution, prepared by dissolving 100 milligrams of bromothymol blue in 100 milliliters of dilute alcohol, and slightly acidify with diluted hydrochloric acid. Add 1.0 N sodium hydroxide solution until the color changes to a clear blue, then boil the solution for 3 minutes agitating it gently during the heating period: a white, crystalline precipitate appears in the liquid. This precipitate is insoluble in 1 N sodium hydroxide solution, but is soluble in dilute hydrochloric acid.

When to a solution of a citrate one tenth its volume of Denigés reagent is added, the mixture heated to boiling, then 0.1 N potassium permanganate solution added, a white precipitate is formed. Denigés reagent is prepared by mixing 5 grams of yellow mercuric oxide with 40 milliliters of water and, while stirring, 20 milliliters of sulfuric acid is added. Finally an additional 40 milliliters of water is added and stirring is continued until solution is complete.

Water—Determine the water content of citric acid by the Karl Fischer method: Anhydrous citric acid contains not more than 0.5 per cent and hydrous citric acid not more than 8.8 per cent of water.

Residue on ignition—Citric acid yields not more than 0.05 per cent of residue on ignition. Weigh accurately 1 to 2 grams of citric acid into a suitable tared crucible. Ignite until thoroughly charred, cool, moisten the residue with 1 milliliter of sulfuric acid and cautiously ignite until the carbon is completely consumed. Conduct the ignition in a place protected from air currents and use as low a temperature as possible to effect the combustion of the carbon. When the carbon has completely disappeared, cool the crucible in a desiccator, and weigh.

Oxalate—Neutralize 10 milliliters of a solution of citric acid (1 in 10) with 10 per cent ammonium hydroxide solution, add 5 drops of dilute hydrochloric acid, cool, and add 2 milliliters of 1 N calcium chloride solution: No turbidity is produced.

Sulfate—To 10 milliliters of a solution of citric acid (1 in 100) add 1 milliliter of 1 N barium chloride solution to which has been added 1 drop of hydrochloric acid: No turbidity is produced.

Heavy metals—Dissolve 2 grams of citric acid in 10 milliliters of water in a 50-milliliter Nessler tube, add I drop of phenolphthalein indicator solution, then add 10 per cent ammonium hydroxide solution until the test solution is faintly pink. Add 2 milliliters of dilute acetic acid and dilute with water to 25 milliliters.

In another 50-milliliter Nessler tube, place sufficient standard lead to give 10 parts per million. Add 2 milliliters of dilute acetic acid and make up to 25 milliliters with water.

Add 10 milliliters of a saturated solution of hydrogen sulfide in cold water to each tube, mix and allow to stand for 10 minutes. The color produced in the citric acid solution should not exceed the 10 parts per million of the standard.

Lead nitrate stock solution—Dissolve 159.8 milligrams of lead nitrate in 100 milliliters of water to which has been added 1 milliliter of nitric acid, then dilute to 1000 milliliters with water. Prepare and store this solution in glass containers free from soluble lead salts.

Standard lead solution—Dilute 10 milliliters of lead nitrate stock solution accurately measured, to 100 milliliters with water. This solution must be freshly prepared. Each milliliter of standard lead solution contains the equivalent of 0.01 milligram of lead. When 0.1 milliliter of standard lead solution is employed to prepare the standard to be compared with a solution of 1 gram of the substance being tested, the comparison solution thus prepared contains the equivalent of 1 part of lead per million parts of the substance tested.

Readily carbonizable substances—Mix 500 milligrams of finely powdered citric acid with 5 milliliters of sulfuric acid in a test tube that has been previously rinsed with sulfuric acid and maintain the temperature of the mixture at 90°C for 1 hour. Transfer the test solution to a comparison tube that is made of colorless glass resistant to the action of sulfuric acid which should contain not less than 94.5 per cent and not more than 95.5 per cent H_2SO_4, determined by titration. The color of the mixture is not darker than that of matching fluid K of the U.S.P.XV.

Assay—Place about 3 grams of citric acid in a tared flask and weigh accurately. Dissolve in 40 milliliters of water and titrate with 1 N sodium hydroxide solution, using phenolphthalein indicator solution (1 gram of phenolphthalein dissolved in 100 milliliters of ethyl alcohol). Each milliliter of 1 N sodium hydroxide solution is equivalent to 64.04 milligrams of $C_6H_8O_7$.

3. TARTARIC ACID

Tartaric acid, HOOC·CHOH·CHOH·COOH, dihydroxysuccinic acid, is a dibasic acid forming four isomers. The common acid is the *dextro* form and is obtained from grapes. In the fermentation process by which wines are made, the tartaric acid is precipitated as a white cake, consisting mainly of the potassium acid salt with some calcium salt. This mixture, known as argols and crude cream of tartar, is treated to change it to calcium tartrate which, in turn, is treated with sulfuric acid to form tartaric acid and insoluble calcium sulfate.

Common tartaric acid occurs in colorless and transparent crystals or as a white powder or granules. It has a strong sour taste. Tartaric acid melts at 168° to 170°C and decomposes at higher temperatures with the formation of an odor resembling that of burned sugar. It has a specific gravity of 1.76 and the specific rotation of a 20 per cent aqueous solution of the *dextro* compound is +11.98°. It is very soluble in water, 139 grams dissolving in 100 milliliters of water at 20°C. It is also soluble in alcohol, 25 grams dissolving in 100 milliliters of absolute alcohol at 15°C.

It has already been noted that tartaric acid occurs in grapes. It is also found in other fruits. Since it is the acid characteristic of grapes it is often used in the beverage industry for the preparation of grape-flavored beverages.

In Table 4-4, the Baumé hydrometer readings, specific gravity, grams per liter, and pounds per gallon corresponding to given percentage tartaric acid solutions are tabulated.

a. National Formulary Requirements

The National Formulary [2] details the following criteria for tartaric acid.

Tartaric acid, dried over sulfuric acid for 3 hours, contains not less than 99.7 per cent of $C_4H_6O_6$.

Description—Tartaric acid occurs as colorless or translucent crystals, or as a white, fine to granular, crystalline powder. It is odorless, has an acid taste, and is stable in air.

TABLE 4-4

Tartaric Acid, Aqueous Solutions
Specific Gravity at 15°C

$C_4H_6O_6$, %	Bé	Sp. Gr.	Grams/liter	Pounds/gallon
1	0.6	1.0045	10.05	0.08383
2	1.3	1.0090	20.18	0.1684
4	2.6	1.0179	40.72	0.3398
6	3.9	1.0273	61.64	0.5144
8	5.2	1.0371	82.97	0.6924
10	6.5	1.0469	104.7	0.8737
12	7.8	1.0565	126.8	1.058
14	9.0	1.0661	149.3	1.246
16	10.3	1.0761	172.2	1.437
18	11.5	1.0865	195.6	1.632
20	12.8	1.0969	219.4	1.831
22	14.0	1.1072	243.6	2.033
24	15.2	1.1175	268.2	2.238
26	16.5	1.1282	293.3	2.448
28	17.7	1.1393	319.0	2.662
30	19.0	1.1505	345.2	2.880
32	20.2	1.1615	371.7	3.102
34	21.3	1.1726	398.7	3.327
36	22.5	1.1840	426.2	3.557
38	23.8	1.1959	454.4	3.793
40	25.0	1.2078	483.1	4.032
42	26.1	1.2198	512.3	4.275
44	27.3	1.2317	541.9	4.523
46	28.5	1.2441	572.3	4.776
48	29.6	1.2568	603.3	5.034
50	30.8	1.2696	634.8	5.298
52	32.0	1.2828	667.1	5.567
54	33.1	1.2961	699.9	5.841
56	34.3	1.3093	733.2	6.119

Solubility—One gram of tartaric acid dissolves in 0.8 milliliter of water and in about 3 milliliters of alcohol. One gram dissolves in 0.5 milliliter of boiling water.

Identification—A: Tartaric acid responds to the tests for tartrates. With a neutral solution of tartrates, 1 N silver nitrate solution produces a white precipitate. On dissolving this precipitate in just sufficient 10 per cent ammonium hydroxide solution and warming, metallic silver is deposited on the side of the test tube, forming a mirror. On adding to a solution of tartaric acid or of a tartrate, acidified with a few drops of acetic acid, a drop of ferrous sulfate solution, prepared by dissolving 8 grams of clear crystals of ferrous sulfate in 100 milliliters of recently boiled and cooled water, then a few drops of 3 per cent hydrogen peroxide solution, and finally an excess of 1 N sodium hydroxide solution, a purplish violet color is produced.

B: Ignite the tartaric acid. It gradually decomposes, emitting an odor resembling that of burning sugar (distinction from citric acid).

Loss on drying—Dry tartaric acid over sulfuric acid for 3 hours: It loses not more than 0.5 per cent of its weight. Conduct the determination on 1 to 2 grams of the tartaric acid previously well mixed and accurately weighed. If the sample is in the form of large crystals, reduce the particle size to about 2 millimeters by quickly crushing. Tare a glass-stoppered, shallow weighing bottle that has been dried for 30 minutes in the same desiccator as that to be used for the sample. Put the sample in the bottle, replace the cover, and weigh the bottle and the charge. By gentle, sidewise shaking, distribute the charge as evenly as practicable to a depth of about 5 millimeters generally, and not over 10 millimeters in the case of a bulky sample. Place the loaded bottle in the desiccator, removing the stopper and leaving it also in the desiccator, and dry the sample for 3 hours. On opening the desiccator, close the bottle promptly and allow it to come to room conditions before weighing in the dried state.

Residue on ignition—Tartaric acid yields not more than 0.05 per cent of residue on ignition. Weigh accurately 1 to 2 grams into a tared crucible of platinum, porcelain, or other suitable material. Ignite until thoroughly charred, cool, moisten the residue with 1 milliliter of sulfuric acid, and cautiously ignite until the carbon is completely consumed. Conduct the ignition in a place protected from air currents and use a temperature as low as possible to effect the combustion of the carbon. When the carbon has completely disappeared, cool the crucible in a desiccator and weigh.

Oxalate—Nearly neutralize 10 milliliters of a solution of tartaric acid (1 in 10) with 10 per cent ammonium hydroxide solution and add 10 milliliters of saturated calcium sulfate solution: No turbidity is produced.

Sulfate—To 10 milliliters of solution of tartaric acid (1 in 100) add 3 drops of hydrochloric acid and 1 milliliter of 1 N barium chloride solution: No turbidity is produced.

Heavy metals—Conduct the test as directed for citric acid.

Other acids—When calculated from the assay, each gram of tartaric acid, previously dried, consumes not less than 13.28 and not more than 13.40 milliliters of 1 N sodium hydroxide solution.

Assay—Place about 2 grams of tartaric acid, previously dried over sulfuric acid for 3 hours and accurately weighed, in an Erlenmeyer flask. Dissolve it in 40 milliliters of water, add a few drops of phenolphthalein indicator solution, and titrate with 1 N sodium hydroxide solution. Each milliliter of 1 N sodium hydroxide solution is equivalent to 75.04 milligrams of $C_4H_6O_6$.

4. MALIC ACID

Malic acid, $HOCH(COOH)CH_2COOH$, L-hydroxysuccinic acid is the fruit acid characteristic of apples and for this reason is also known as apple acid. It is a white or colorless crystalline solid, melting at 99° to 100°C and boiling at 140°C with some decomposition. It has a specific gravity of 1.595. Malic acid is very soluble in water and in alcohol.

Malic acid is found in apples, peaches, cherries, grapes, pineapple, and other fruits and in celery and rhubarb stalks. It is manufactured synthetically by the hydration of maleic acid prepared by the catalytic oxidation of benzene vapor. Malic acid was isolated as long ago as 1785 by Scheele.

Malic acid is recommended for use in apple, cherry, and celery-flavored carbonated beverages. The price of acids used for the manufacture of carbonated beverages in drum, carload, and tank car lots shows why the use of malic acid is economically disadvantageous. Thus for a number of years (1952–56) technical malic acid sold for about 50 cents per pound; anhydrous citric acid, U.S.P., 25 to 30 cents per pound and hydrous citric acid 27 to 27.5 cents per pound; edible lactic acid, 50 per cent, 17 to 18 cents per pound and 80 per cent lactic acid, 30 to 31 cents per pound; phosphoric acid, food grade, 75 per cent, 5 to 5.5 cents per pound and phosphoric acid, N.F. 85 per cent, 8 (1952) to 6.4 (1956) cents per pound.

5. LACTIC ACID

Commercial lactic acid is sold in several grades and concentrations, such as edible 50 per cent, edible 80 per cent, and U.S.P. 85 per cent. The last is a sirupy, hygroscopic liquid with a specific gravity of about 1.2. It contains 85 to 90 per cent lactic acid of which about 10 to 15 per cent is lactic anhydride, for when lactic acid is concentrated to greater than 50 per cent it is partially converted into lactic anhydride. It cannot boil or be distilled at atmospheric pressure without decomposition. It boils under reduced pressure (15 mm) at 122°C by which means it can be distilled and purified. Lactic acid is miscible with water, alcohol, and glycerol. It is made on a commercial basis by fermentation of sugars and cornstarch.

Lactic acid, ordinary lactic acid, racemic DL-lactic acid, 2-hydroxypropionic acid, $CH_3CH(OH)COOH$, has a melting point of 16° to 18°C and a specific gravity of 1.249 when pure. It is widely used as an acidulant in the manufacture of carbonated beverages.

Edible 50 per cent lactic acid has a specific gravity of 1.135 and 100 grams of this grade is equivalent to 88.5 milliliters. Edible 80 per cent lactic acid has a specific gravity of 1.205 and 100 grams of this grade is equivalent to 83.5 milliliters.

a. United States Pharmacopoeia Requirements [1]

Lactic acid is a mixture of lactic acid ($C_3H_6O_3$) and lactic anhydride ($C_6H_{10}O_5$), equivalent to a total of not less than 85 per cent, and not more than 90 per cent, by weight, of $C_3H_6O_3$.

Description—Lactic acid is a colorless or yellowish, nearly odorless, sirupy liquid. It absorbs water on exposure to moist air and cannot be distilled under normal pressure without decomposition. The specific gravity of lactic acid is about 1.206.

Solubility—Lactic acid is miscible with water, with alcohol, and with ether. It is insoluble in chloroform.

Identification—Lactic acid responds to the test for lactate. When solutions of lactates are acidified with sulfuric acid, 0.1 N potassium permanganate solution is added, and the mixture heated, acetaldehyde, recognizable by its distinctive odor, is evolved.

Residue on ignition—A 5 milliliter portion of lactic acid yields not more than 6 milligrams of residue on ignition (0.1 per cent).

Chloride—To 10 milliliters of a solution of lactic acid (1 in 100), acidified with nitric acid, add a few drops of 1 N silver nitrate solution: No opalescence is produced immediately.

Citric, oxalic, phosphoric, or tartaric acid—To 10 milliliters of a solution of lactic acid (1 in 10) add 40 milliliters of 0.04 N calcium hydroxide solution and boil for 2 minutes: No turbidity is produced.

Sulfate—To 10 milliliters of a solution of lactic acid (1 in 100) add 2 drops of hydrochloric acid and 1 milliliter of 1 N barium chloride solution: No turbidity is produced.

Heavy metals—Dilute 1.7 milliliters (2 grams) of lactic acid with 10 milliliters of water and add 1 drop of phenolphthalein indicator solution, followed by 10 per cent ammonium hydroxide solution until the solution is faintly pink. Add 2.5 milliliters of dilute hydrochloric acid and sufficient water to make 25 milliliters. Conduct the test as directed under citric acid. The heavy-metals limit for lactic acid is 10 parts per million.

Sugars—To 10 milliliters of hot Fehling's solution add 5 drops of lactic acid: No red precipitate is formed.

Readily carbonizable substances—Carefully superimpose 5 milliliters of lactic acid on an equal volume of sulfuric acid in a test tube and keep the temperature at 15°C: No dark color develops at the zone of contact of the two liquids within 15 minutes.

Assay—To about 2.5 milliliters of lactic acid, accurately weighed in a tared 250-milliliter flask, add 50 milliliters of 1 N sodium hydroxide solution and boil the mixture for 20 minutes. Add phenolphthalein indicator solution and titrate the excess alkali in the hot solution with 1 N sulfuric acid. Perform a blank determination with the same quantities of the same reagents and in the same manner. Each milliliter of 1 N sodium hydroxide solution is equivalent to 90.08 milligrams of $C_3H_6O_3$.

6. ACETIC ACID

Acetic acid, CH_3COOH, is one of the oldest acids known to man for it is the main acid of vinegar and thus has been known for centuries. It is a colorless liquid with a pungent odor which, on dilution, has a characteristic vinegarlike odor and a sour taste.

Glacial acetic acid boils at 118°C, solidifies at 16.7°C; it has a specific gravity of 1.047 to 1.050 and a refractive index of 1.3718. Acetic acid is miscible with water and with alcohol. Care must be taken in its storage since it expands on solidification and its solidification temperature is relatively high, namely, 16.7°C.

Acetic acid is prepared from pyroligneous acid, obtained from the destructive distillation of wood, by neutralization with lime and precipitation as calcium acetate, which may be subsequently decomposed with sulfuric acid to yield acetic acid. Its complete synthesis depends on the formation of acetaldehyde from acetylene and water and the subsequent oxidation by air of the acetaldehyde to acetic acid. It can also be prepared by the oxidation of ethyl alcohol. Acetic acid is available in a number of grades and percentages, such as 99.5, 90, and 80 per cent.

Kühles [3] mentions that in times of stringency, acetic acid can be used as a carbonated beverage acidulant. He suggests that the 80 per cent acid is diluted with three times its volume of water to form a 20 per cent solution and that this will have an acid strength equivalent to a 50 per cent citric acid solution. He also suggests that one third of the customary acidulant can be replaced by acetic acid without noticeable effect on the flavor. Higher amounts will cause the vinegar character to appear but this can be offset by use of 100 grams of sodium chloride to 26.5 gallons of the added flavor. It must also be remembered that when acetic acid is used as part of a flavor it adds to the total acidity.

7. PHOSPHORIC ACID

Phosphoric acid, orthophosphoric acid, is a colorless, odorless, sirupy liquid containing about 85 per cent H_3PO_4. It has a specific gravity of about 1.7. It is miscible with water and alcohol.

Phosphoric acid is prepared commercially by burning elementary phosphorus in air with subsequent hydration of the phosphorus pentoxide, P_2O_5, originally formed.

Phosphoric acid is widely used as an acidulant of carbonated beverages particularly of kolas and root beers and since these constitute, as noted in Chapter 1, a substantial portion of all carbonated beverages sold, phosphoric acid is possibly the most important soft-drink acid. It is the least expensive of all of the acids used in carbonated beverages, as noted in the section on malic acid. It has the greatest relative strength among the acids used for this purpose. It is not suitable for all types of soft drinks and for this reason, even though the others are more expensive, phosphoric acid has not replaced them.

TABLE 4-5

Specific Gravity of Aqueous Phosphoric Acid

Bé	Sp. Gr.	H_3PO_4, %	Grams/ liter	Pounds/ gallon
0.6	1.0038	1	10.04	0.0838
1.3	1.0092	2	20.18	0.1684
2.8	1.0200	4	40.80	0.3405
4.3	1.0309	6	61.85	0.5162
5.8	1.0420	8	83.36	0.6957
7.3	1.0532	10	105.3	0.8789
8.8	1.0647	12	127.8	1.066
10.3	1.0764	14	150.7	1.258
11.8	1.0884	16	174.1	1.453
13.3	1.1008	18	198.1	1.654
14.8	1.1134	20	222.7	1.858
16.3	1.1263	22	247.8	2.068
17.8	1.1395	24	273.5	2.282
19.2	1.1529	26	299.8	2.501
20.7	1.1665	28	326.6	2.726
22.2	1.1805	30	354.2	2.955
25.8	1.2160	35	425.6	3.552
29.4	1.2540	40	501.6	4.186
32.9	1.2930	45	581.9	4.856
36.4	1.3350	50	667.5	5.570
39.9	1.3790	55	758.5	6.329
43.3	1.4260	60	855.6	7.140
46.7	1.4750	65	958.8	8.001
50.0	1.5260	70	1068.0	8.914
53.2	1.5790	75	1184.0	9.883
56.2	1.6330	80	1306.0	10.900
59.2	1.6890	85	1436.0	11.980

a. United States Pharmacopoeia Requirements

The United States Pharmacopoeia [1] details the following criteria for phosphoric acid, U.S.P.XV.

Phosphoric acid is a colorless, odorless liquid of sirupy consistency. It is miscible with water and alcohol.

Assay—Weigh accurately about 1 gram in a glass-stoppered flask, dilute with about 100 milliliters of water, add 0.5 milliliter of thymolphthalein indicator solution, and titrate with 1 N sodium hydroxide solution: Each milliliter of sodium hydroxide solution consumed is equivalent to 49.0 milligrams of H_3PO_4. Not less than 85 per cent is found.

Chloride—Two milliliters show not more than 0.01 milligram of Cl (0.0003 per cent). Standard chloride solution—Dissolve 165.0 milligrams of dried sodium chloride in sufficient water to make exactly 1000 milliliters. This solution contains the equivalent of 0.10 milligram of chlorine (Cl) in each milliliter.

Nitrate—Dilute 2 milliliters with water to 10 milliliters and add 5 milligrams of sodium chloride, 0.1 milliliter of indigo carmine indicator solution and 10 milliliters of sulfuric acid: The blue color is not entirely discharged within 5 minutes (about 0.0005 per cent).

Sulfate—Dilute 12 milliliters with 190 milliliters of water, bring the solution to a boil, add 10 milliliters of 1 N barium chloride solution and allow to stand overnight. If a precipitate is present, filter, wash, ignite, and weigh: The residue weighs not more than 1.5 milligrams more than that obtained in a blank (0.003 per cent).

Reducing substances—Add 5 milliliters of 0.1 N bromine solution to each of two 500-milliliter iodine flasks. To each flask add 120 milliliters of cold dilute sulfuric acid (1 in 60) and to one of the flasks add, rapidly, 10 milliliters of sample. Stopper the flasks, shake gently, seal with water, and allow to stand at room temperature for 10 minutes. Cool in an ice bath, add 10 milliliters of potassium iodide solution, and reseal with water. Shake gently to dissolve the fumes and place in the ice bath for about 5 minutes. Wash the stoppers and inside walls of the flasks and titrate each with 0.01 N sodium thiosulfate solution, adding starch indicator solution near the end of the titration: The difference between the two titrations does not exceed 1.0 milliliter of the 0.01 N sodium thiosulfate solution.

Volatile acids—Dilute 37.5 milliliters with 75 milliliters of freshly boiled and cooled water in a distilling flask provided with a spray trap and distill 50 milliliters. To the distillate, add phenolphthalein indicator solution and titrate with 0.1 N sodium hydroxide solution: Not more than 0.1 milliliter of the sodium hydroxide solution is required for neutralization (0.001 per cent as acetic acid).

Alkali and other phosphates—Dilute 1.8 milliliters with 100 milliliters of water, add, with stirring, a solution of 15 grams of lead acetate in 50 milliliters of water, and dilute with water to 200 milliliters. Filter and pass hydrogen sulfide through 100 milliliters of the filtrate until the lead is completely precipitated. Filter, wash with 20 milliliters of water, add to the filtrate 2 drops of sulfuric acid, evaporate to dryness, ignite gently, and weigh: The residue weighs not more than 3 milligrams more than that obtained in a blank (0.2 per cent).

Arsenic—Test 3.0 grams: The stain produced is not darker than that produced by 0.003 milligram of arsenic (0.0001 per cent). Select all the reagents used in this test to have as low a content of arsenic as possible so that a blank test results in either no stain or one that is barely discernible.

Heavy metals—Dilute 2.4 milliliters with 10 milliliters of water, add 12 milliliters of dilute ammonium hydroxide solution (1 in 5), dilute to 40 milliliters, and to 30 milliliters of the resulting solution, add 10 milliliters of saturated cold hydrogen sulfide solution: Any brown color produced is not darker than that of a control containing 0.02 milligram of added lead and the remaining 10 milliliters of the solution (0.001 per cent).

Iron—Dilute 6 milliliters to 100 milliliters. To 5 milliliters of the dilution, add 8 milliliters of 0.5 N sodium acetate solution, 6 milliliters of hydroxylamine hydrochloride solution (1 in 10), 4 milliliters of a slightly acidified solution of orthophenanthroline (1 in 1000), and dilute to 25 milliliters: Any red color produced within 1 hour is not darker than that of a control containing 0.015 milligram of iron (0.003 per cent).

8. ACIDULATION

The most important factor in the choice of an acid for a given carbonated beverage flavor is its ability to bring out and enhance the flavor in question. On this basis, tartaric acid is the best acid for a grape flavor, citric acid for the citrus flavors, and malic acid for apple flavor. Actually, however, other factors, such as price and availability, enter into the picture of the choice of an acid. Commonly, as mentioned, phosphoric acid is used as the acidulant of choice for the kolas, root beers, sarsaparilla, and similar beverages; citric acid is the acid of choice for fruit flavors, particularly for citrus fruit flavors and it is also the acid of choice for ginger ales; tartaric acid is the acid of choice for grape flavor; and malic acid is used for apple, celery, and cherry flavors.

The concentration of the acids commonly used in carbonated beverages and the pH of these soft drinks has been given in Table 4-1 in connection with the discussion on pH in this chapter.

Acidulation is, however, not only important as a flavor factor, but has other functions as well. The more important of these are: First, acids assist in preserving the sirup and beverage by killing microorganisms and making the environment, that is, the beverage, unsuitable for their growth; second, acids catalyze the inversion of sucrose to invert sugar as has been explained in Chapter 2; and third, acids convert sodium benzoate to benzoic acid which exerts its preservative action only in an acid environment.

The acids used for beverages vary in strength. This is due to their structure and their degree of ionization. Thus lactic acid, $CH_3CH(OH)COOH$, is a monobasic acid, malic acid, $HOOCCH_2CH(OH)COOH$, and tartaric acid, $HOOCCH(OH)\cdot CH(OH)COOH$, are dibasic acids, and citric acid, $HOOHCH_2COH(COOH)CH_2COOH$ and phosphoric acid H_3PO_4 are tribasic acids. The degree of dissociation and the ionization constants of these acids were tabulated in Table 4-2. In Table 4-6, the relative strengths of these acids are tabulated.

TABLE 4-6

Relative Strengths of Beverage Acids [a]

Acid	Molecular weight	Equivalent weight	Relative strength
Lactic	90	90	100
Citric	210	70	125
Tartaric	150	75	200
Phosphoric	98	49 [b]	650

[a] After M. Sivetz, in Bottlers' and Glass Packers' Handbook, p. 29, New York, Food Engineering, McGraw-Hill.
[b] Considered as a dibasic acid.

TABLE 4-7

Beverage Acid Equivalents [a]

100 grams of stated acid	is equivalent to:	grams	milliliters
Citric acid, hydrate	tartaric acid	107	
	lactic acid, anhyd.	129	
	lactic acid, 80%	160	133.5
	lactic acid, 50%	257	227.5
	acetic acid, 80%	50	
Tartaric acid	citric acid, hydrate	93	
	lactic acid, anhyd.	120	
	lactic acid, 80%	149	124.5
	lactic acid, 50%	240	212.5
	acetic acid, 80%	47	
Lactic acid, anhydrous	citric acid, hydrate	77.5	
	tartaric acid	83	
	lactic acid, 80%	123	
	lactic acid, 50%	200	
	acetic acid, 80%	39	
Lactic acid, 80% 100 g 83.5 ml	citric acid, hydrate	62	
	tartaric acid	67	
	lactic acid, anhyd.	80	
	lactic acid, 50%	160	141.5
	acetic acid, 80%	31	
Lactic acid, 50% 100 g 88.5 ml	citric acid, hydrate	38.8	
	tartaric acid	41.5	
	lactic acid, anhyd.	50	
	lactic acid, 80%	62	51.7
	acetic acid, 80%	19	
Acetic acid, 80%	citric acid	200	
	tartaric acid	214	
	lactic acid, anhyd.	248	
	lactic acid, 80%	320	
	lactic acid, 50%	514	

[a] After R. Kühles, *Handbuch der Mineralwasser-Industrie*, Lübeck, Antäus-Verlag, 1947.

It is clear that phosphoric acid is the strongest acid of this group and that lactic acid is the weakest. This means that for a given weight, phosphoric acid will contribute more acidity. Phosphoric acid will be able to produce the lowest pH of this group of acids. It must be stressed again that sourness alone is not the controlling factor in the selection of an acid for a given beverage flavor.

LITERATURE CITED

1. *The Pharmacopeia of the United States of America* (The United States Pharmacopeia), *Fifteenth Revision,* U.S.P. XV, U.S. Pharmacopeial Convention Inc., Washington; Easton, Pa., Mack Publishing, December 1955.
2. *The National Formulary,* 10th Ed., N.F.X. Am. Pharm. Assoc., Philadelphia, Lippincott, 1955.
3. R. Kühles, *Handbuch der Mineralwasser-Industrie,* Lübeck, Antäus-Verlag, 1947.

SELECTED REFERENCES

M. B. Jacobs, *Synthetic Food Adjuncts,* New York, Van Nostrand, 1947.

M. B. Jacobs, Ed., *Chemistry and Technology of Food and Food Products,* 2nd Ed., New York, Interscience Publishers, 1951.

M. Sivetz in *Bottlers' and Glass Packers' Handbook,* p. 29, New York, Food Engineering, McGraw-Hill.

Chapter 5

WATER AND WATER TREATMENT

The aqueous portion of a carbonated nonalcoholic beverage makes up the bulk of the final product. For this reason, it must be considered a most important ingredient. In general, the water used to manufacture the beverages we are discussing should meet the sanitary requirements of the Public Health Service, United States Department of Health, Education, and Welfare.[1]

1. CRITERIA

To meet such standards, the water should be clear, colorless, odorless, tasteless, free of iron and manganese, and of acceptable bacteriological quality. The general requirements for water for food purposes are (1) the turbidity should not exceed 10 ppm on the silica scale to conform with United States Public Health Service *Drinking Water Standards;* (2) the color should not exceed 20 ppm on the standard cobalt scale to conform with United States Public Health Service *Drinking Water Standards;* (3) the water should have no odor; (4) the water should have no objectionable taste; (5) the manganese content should not be over 0.1 ppm; (6) the iron content should not be over 0.1 ppm; and (7) the bacteriological quality of all the water used throughout the plant should meet the requirements for drinking water.

More specifically, the requirements of water for the manufacture of carbonated beverages are that the water must be clear, colorless, free from objectionable tastes, free from objectionable odors, free from iron, free from organic matter, low in alkalinity, and sterile. Each of these factors will be considered.

2. CLASSIFICATION

The sources of water used for the manufacture of soft drinks can be classified as (1) surface waters, (2) groundwaters, and (3) rainwater. The first group can be further subdivided as (a) streams including brooks, creeks, and rivers, which are often impounded to form (b) reservoirs, and (c) ponds and (d) lakes. The subdivision of groundwaters can be classified as (a) springs, (b) deep wells, and (c) shallow wells.

The various water sources may be arranged in the order of their normal clarity before treatment as follows: (1) deep wells, (2) shallow wells, (3) springs, (4) rainwater, (5) reservoirs, (6) lakes, (7) streams, and (8) ponds. It is to be noted that if a water contains iron compounds in excess of 1 ppm, when this water is subsequently exposed to air, the iron will oxidize yielding a cloudy water.

Waters can also be arranged according to their mineral content. Rainwater will, since it can only absorb a limited amount of ions and gases from the air as it falls, have the least amount of mineral matter. Surface water will contain considerably more mineral matter than rainwater, but usually less than groundwaters. Generally the water from deep wells will have the highest mineral content of all water sources.

3. WATER REQUIREMENTS

It is seldom possible for the beverage manufacturer to obtain a source of water supply which will be adequate in every respect with regard to the criteria we have set up for beverage water. Consequently the water used must be treated in some manner to make it suitable.

In beverage bottling, the customary arrangement of the water treatment equipment often comprises (1) chemical feeds for ferrous sulfate (copperas) or alum, soda ash, lime, and hypochlorite, (2) cold-process lime water softener; (3) sand filters, (4) activated carbon filters, and (5) polishing filter. Many carbonated beverage manufacturers use a coagulant feed, sand filter, and activated-carbon filter even on a municipal water supply to avoid any possibility of inadvertently incorporating into their beverages the turbidity, corrosion deposits, and off-tastes and off-odors that occasionally get into municipal waters.

An equivalent water-treatment system is considered in the section of the methods for the production of clear water.

Other methods, such as the use of organic cation exchangers, may also be employed by the beverage manufacturer, particularly for the production of water of low alkalinity. These methods can also be used for making water of low hardness.

It is generally not necessary for the manufacturer of soft drinks to produce a water of low or zero hardness for the beverage itself, consequently, there is no need to go to the expense of purchasing equipment or chemicals for this purpose. It is, however, advisable to have soft water for bottle washing. Therefore, where such water is not available as a municipal supply, it is advantageous for the beverage manufacturer to have the portion of his water used for bottle washing softened. One must also remember that too much hardness is undesirable and will cause precipitations of certain dyes used for the coloring of carbonated beverages. This is discussed in Chapter 9.

a. Clear Water

Clear water is essential for the soda-water manufacturer for first, it will enhance the appearance of the final product; second, nuclei for the liberation of dissolved carbon dioxide will not be present; and third reactions with coloring and flavoring matter in the beverage will be avoided.

Nearly all waters, particularly surface waters, as noted previously, have various degrees of turbidity caused by the dispersion of colloidal particles in the water picked up as it flows over or through the ground. Such colloidal particles, whether organic or inorganic, cannot be filtered out by ordinary methods in which bag or sand filters are employed. These colloidal particles must first be coagulated so that they can be filtered out of the water.

Generally soda ash or lime and copperas (ferrous sulfate) or alum (potassium aluminum sulfate) are used to coagulate the colloidal matter with the principal part of the filtration being done by a blanket formed by the floc. In actual practice, the mixture of chemicals and water is passed upward in some types of equipment through a tank with an increasing cross-sectional area after thorough mixing and arrangement of adequate detention time. This will assist in coagulating all the matter causing the turbidity to form larger particles. Normally these particles would settle out if the water were still. Since, however, the water is flowing and rising upward and since the cross-sectional area of the treatment tank is increasing, a section of the tank will be reached where the upward flow rate of the water will be reduced sufficiently so that the lifting effect on the agglomerated particles will be balanced by the sedimentation rate of the same agglomerated particles. The coagulated floc remains in this section of the tank performing the preliminary filtration.

The water obtained by this treatment will still not be adequately clear and will require additional handling. Some of the coagulated particles may be carried past the coagulum floc. These can be removed and the clarity desired obtained by passing the water through an additional series of steps comprising a sand or calcite filter, an activated-carbon section, and a disc filter. The last is required to give the water a final polishing if a beverage of exceptional sparkling character is desired.

b. Colorless Water

To insure uniformity of color from batch to batch in the manufacture of carbonated beverages and to avoid interference with the added color or subsequent precipitation of either the added color or the undesirable color in the water, it is best to start with a colorless water. Color in a water generally is indicative of the presence of organic matter. This may precipitate out of the beverage on standing. It may also cause off-odor or off-taste.

Surface waters are often colored because they may pick up organic or vegetative matter or be contaminated with swamp drainage. Groundwaters, however, are seldom colored.

Soluble or colloidally dispersed color and color from swamp drainage may be eliminated by oxidation with chlorine or its compounds as detailed in the subsection on sterilization. The color attributable to vegetative or organic matter may be removed by the coagulation and sedimentation process described for the removal of turbidity in water.

c. Tasteless Water

The water to be used for the manufacture of soda pop and soft drinks must be free of foreign tastes or any pronounced taste, for if this condition is not fulfilled, it becomes increasingly difficult to control the flavor of the drink as each batch is made. Possibly the most important factor in the taste of water as supplied is that of high alkalinity. The control of alkalinity will be discussed in a subsequent section. The presence of iron will also affect adversely the taste of the water. Methods for the removal of iron are also considered in a subsequent section.

Still another factor to consider in connection with the taste of the water to be used is the application of chlorine for sterilization. Excess chlorine produces an off-taste in the water and thus must be eliminated. It must also be remembered that chlorine may react with phenolic impurities in the water to form even more unpleasant-tasting and malodorous chlorophenols. These compounds can be removed by the use of an activated-carbon filter.

d. Odorless Water

The water to be used for the manufacture of carbonated beverages must be free from unpleasant odors. The reasons for this are the same or virtually the same as those mentioned in connection with the need for tasteless water.

Among the substances which cause unpleasant odors in waters are hydrogen sulfide and other sulfides, phenols, and the chlorophenols already mentioned. The most common method for the removal of hydrogen sulfide in water used for bottling is chlorination, but at times aeration can also be used. Excess chlorine and any malodorous chlorophenols may be eliminated by the use of an activated-carbon filter.

e. Iron

It is essential that water for carbonated beverages is free from iron. Indeed, the requirement for limitation of iron in water for carbonated beverages is more stringent than that for potable water. Thus the specific requirement for iron-free water for carbonated beverages is water containing less than 0.1 ppm of iron whereas the United States Public Health Service *Drinking Water Standards* permits a total of 0.3 ppm of iron and manganese.

This limitation is necessary on several counts. First, iron may precipitate as one of its salts or compounds in the bottle, forming an unsightly deposit or a yellowish to reddish ring of ferric hydroxide in the bottle; second, it may react with and precipitate a lake of the coloring matter used in the beverage; third, it may affect the taste of the beverage either by its presence or by reaction with one of the flavoring components.

Iron is present in water obtained from deep wells usually as ferrous bicarbonate. When such waters are exposed to air or are aerated, the ferrous bicarbonate is oxidized to form a ferric compound and precipitates as ferric hydroxide. It can readily be seen that such raw waters cannot be used for beverages.

The three principal methods for the removal of iron are (1) aeration for oxidation, with subsequent settling and filtration; (2) manganese zeolite treatment; and (3) less frequently used by the beverage manufacturer, since elimination of hardness is not too important, cation exchange.

Oxidation of deep well water can be performed at times by pumping the water to an open storage tank and allowing it to stand before sending it to a precipitator. In one type of water-treatment equipment, such water is permitted to fall through the air before it enters the precipitator.

Oxidation of ferrous ion to ferric ion goes more slowly in acid solution. The cold-lime or lime-soda processes of water treatment assist in precipitating the iron by raising the pH. In some types of equipment, an aerator is used before the settling tank or precipitator.

f. **Organic Matter**

The presence of organic matter in water to be used for bottling is objectionable because it will cause unnecessary foaming or boiling at the filler and gushing and thus loss of carbon dioxide when the bottle of beverage is opened; it may make the water a suitable medium for the growth of microorganisms, including possibly harmful organisms; and the beverage may be considered spoiled because such organic matter in the water may have been extracted from deposits of dead animal and plant matter and from human, animal, or industrial wastes. Microorganisms, such as bacteria, yeasts, molds, algae, and protozoa, may constitute or contribute to the organic matter present in raw waters.

Some of this organic matter, if present, can be removed by the coagulation and filtration method described in the subsection relating to the preparation of clear water. Most of the remainder of such organic matter is made harmless or destroyed by the calcium hypochlorite or other chlorinating or sanitizing agent used in the sterilization process. This is discussed in greater detail in subsequent sections.

g. **Low-Alkalinity Water**

An important step in the preparation of adequate water for beverages is the production and maintenance of a water supply with low alkalinity. This is considered by some to be a matter of paramount importance. Usually 80 ppm is considered to be the upper permissible limit for water to be used for carbonated beverages. This is, however, an arbitrary value and not all bottlers follow it, for some set a standard as high as 110 ppm and others insist on an alkalinity as low as 20 ppm. Since some waters have alkalinities as high as 500 ppm or higher, a value of even 110 ppm implies the need of a substantial reduction in alkalinity.

The four principal reasons for reducing the alkalinity of water to be used for beverages are (1) production of optimum flavor and tang in the beverages; (2) to guard against spoilage by maintaining the correct acidity and not neutralizing the acidity of the beverage sirup; (3) elimination of off-tastes attributable to alkaline salts; and (4) to produce a beverage of uniform taste, flavor, and quality.

The main objection to water of high alkalinity is that it may neutralize a portion of the acid in acid-type soft drinks and thus yield a flat drink, that is a beverage without sufficient tang. The taste of such drinks is lacking in sharpness and acidity.

Since, as has been stressed in Chapter 4 on acids and acidulation, the acid component of a carbonated beverage helps in producing and keeping a sterile beverage, the use of water of high alkalinity for beverages will counteract this effect by neutralizing a portion of the acid thus reducing not only the total titratable acidity but also the acidity expressed as pH.

An additional objection to the off-tastes caused by the use of water of high alkalinity is the fact that such tastes tend to linger. This, as discussed in Chapter 6 on flavors and flavoring, is undesirable.

Uniformity of product is a criterion that every beverage manufacturer seeks to achieve. One of the problems encountered with waters of high alkalinity is that this alkalinity is not constant and may vary particularly from season to season. For this reason, the manufacturer cannot rely on such water without treating it to obtain a water of uniform alkalinity.

The reduction af alkalinity is usually accomplished by use of lime to precipitate calcium as calcium carbonate and the use of ferrous sulfate as a coagulant in the cold lime water softner. Calcium hypochlorite is used as the sterilizing agent.

h. **Sterile Water**

There are a number of methods by which water may be sterilized, such as special distillation methods, exposure to ultraviolet rays or, in more modern techniques, other types of radiation, such as gamma rays, electron beams, etc., sonic vibration, ultrafiltration, and related types of filtration, but the method of choice for beverage manufacture is the use of compounds capable of yielding free chlorine. Chlorine, itself, may have been used and organic compounds yielding chlorine, such as Chloramine-T, Dichloramine-T, Halozone, azochloroamide, etc., could possibly be used, but the most commonly used compound is calcium hypochlorite. This is sold commercially in various grades, such as chlorinated lime or bleaching powder, $Ca(ClO)_2 \cdot CaCl_2$, which contains about 35 per cent of active chlorine or high-test calcium hypochlorite, often sold under trade names, such as Perchloron or H.T.H., containing about 50 per cent of available chlorine.

Mention has been made in the section on tasteless water of the need to remove excess chlorine to prevent off-tastes. This is stressed again. It is done by passing the water through an activated-carbon filter. Activated carbon has the property of adsorbing, that is, collecting on its surfaces, certain vapors, odorous and other organic compounds, and colors and thus removing them from solutions or from gaseous mixtures. This property is relatively selective, for compounds of higher molecular weight are generally adsorbed more strongly than those of lower molecular weight.

Activated carbons vary in properties, some being better adsorbents than others. Usually the activated carbons sold for water purification are adequate for the removal of chlorine and chlorophenols, but the completeness of removal is a function of the concentration of the chlorine in the untreated water, the type of carbon used, and the length of time the carbon was used.

4. WATER TREATMENT

As mentioned, there are a number of different types of equipment commercially available for the purification of water for carbonated beverages. Some description of the treatment required to obtain certain objectives has been given in the preceding sections of this chapter. A description of three of these types of equipment, such as the Permutit Precipitator,[2] the "Oxylator" system,[3] and cation exchange [4] should be helpful in an understanding of the entire purification cycle.

a. Permutit Precipitator

The Permutit equipment can be used for water purification by either coagulation alone or by both coagulation and alkalinity-reducing treatments, depending, of course, on the water source and the treatment required. There is no change in the equipment itself; only the chemicals that are to be added are changed. The apparatus of Figure 2, comprises chemical tanks, the coagulator-dealkalizer itself, a pump, a calcite filter, and an activated-carbon adsorber.

(1) Chemical Feeds

Two chemical tanks, lined with a baked-on plastic enamel that is unaffected by the chemicals used and that will not add any taste or odor to the water, are used, one of which is equipped with a motor-driven stirrer. The chemical tank having the stirrer is employed for the preparation of those chemical mixtures which are fed to the precipitator as slurries or suspensions, such as lime. The other chemical tank is used for the addition of chemicals which form clear solutions, such as the common coagulants ferrous sulfate and alum.

(2) Precipitation Tank

The precipitation tank consists of (1) a mixing-coagulation zone, (2) a sludge-filter zone, and (3) a clear-water zone. Raw water

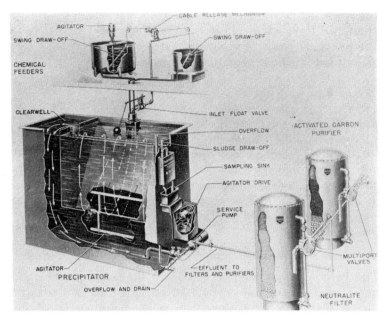

FIGURE 2. Permutit Precipitator, Neutralite Filter, and Activated Carbon Purifier (Courtesy of the Permutit Co., a division of Pfaudler Permutit Inc., New York, N. Y.)

enters the top of the mixing-coagulation zone, flows downward to enter the bottom of the sludge-filter zone, and then upward through the sludge-filter zone to the clear-water zone. The partially treated water is drawn off near the bottom of the clear-water zone.

In the mixing-coagulation zone the chemicals required for proper treatment, fed from the two chemical tanks, are mixed with the raw water to obtain optimum flocculation. A horizontal stirrer gives a slow, rolling motion to the water and the added chemicals. Sufficient time is allowed for the water to remain in this zone so that the chemicals can react completely and flocculate.

The flocs formed in the mixing-coagulation zone are filtered through a suspended sludge blanket which is maintained in its position by the upward but constantly decreasing velocity of the partially treated water. The contact with the previously formed sludge gives additional assurance that the added chemicals will react completely. The sludge blanket serves to entrain the substances causing the turbidity, the organic matter, and the precipitates of the chemicals. The suspended sludge blanket forms a line of demarcation, the water passing through the sludge being clear. This partially treated water passes into the clear-well zone.

The water in the clear-well zone is used not only to give the purified water required for making the carbonated beverages, but also serves to backwash the calcite filter and the activated-carbon adsorber. There is generally sufficient water available in the clear-water zone for this purpose. There is also, generally, sufficient water available in the clear-water zone to smooth out the variations in flow caused by stopping and starting the bottle-filling machine and for adequate water at the pump.

(3) Calcite Filter

The partially treated water is pumped from the clear-water zone to the calcite filter. In the Permutit equipment, a granular filtering medium which they term "Neutralite" is used. This filter serves to remove all of the fine floc particles which pass the sludge blanket and renders the slightly acidic water noncorrosive by making the effluent water slightly alkaline. The principal function of this filter with waters that have been treated to reduce their alkalinity is to filter out the fine flocculent particles.

The calcite filter is a closed, cylindrical, corrosion-resistant steel shell. In the dished bottom of the calcite filter, $\frac{1}{4}$ to $\frac{1}{2}$ inch gravel is placed. On top of this is a 5-inch layer of $\frac{1}{8}$ to $\frac{1}{4}$ inch gravel and on top of the smaller-sized gravel there is a 23-inch layer of crushed calcite. The partially treated water from the clear-water zone of the precipitator enters at the top and percolates downward through the crushed calcite and the gravel beds.

(4) Activated-Carbon Adsorber

After filtration through the calcite bed, the water is passed through an activated-carbon adsorber which is similar in construction to the calcite filter. It contains $\frac{1}{4}$ to $\frac{1}{2}$ inch gravel in the dished bottom, on top of which is a 5-inch layer of $\frac{1}{8}$ to $\frac{1}{4}$ inch gravel then a 3-inch layer of coarse sand, and finally a 27-inch layer of activated carbon. The Permutit system utilizes a special type of activated carbon. The passage of the water through the activated carbon adsorber removes taste, odor, and chlorine and the water goes to the cooler of the carbonator. Here, too, the water enters at the top of the carbon layer.

(5) Operation

In starting the system, the charge of chemicals is placed in the chemical feed tanks and the feeding device is started. The flow of water is started through the precipitator, but at half the rated capacity and the filter and adsorber are by-passed. The system is allowed to operate until a sludge is formed and after the sludge blanket is formed the chemical feeds are adjusted as required.

If the bottling plant and the water system are shut down each night, it is necessary to start the water system the following day at least 1 hour before the full demand. The agitator is started first to stir up the sludge which had settled to the bottom overnight and stirring is continued for 10 to 15 minutes before starting the flow of water. As in the case of an initial start-up, water is run in at half the rated capacity and then the flow is gradually increased to full. These changes in flow rate are made at the inlet rate-set valve. If there is no immediate demand for water, the filter and adsorber can be by-passed to spare them.

(a) Precautions

In order to obtain proper coagulation it is necessary to control the pH, preferably at a value of 6.5 to 7.5 if alum is used and 10.2 if ferrous sulfate (copperas) is used. In most systems copperas is used, because the range of iron precipitation is 4.5 to 11 whereas that of alum is between 5.5 and 7.5. Indeed copperas displaced alum as the coagulant of choice in the carbonated-beverage industry, because there was less chance that iron would fail to be precipitated than alum. If any alum or iron is not precipitated in the coagulation tank, it is likely to pass into the water used for carbonation. This is serious since the hydroxides may precipitate after the beverage is bottled. The amount of adjustment depends on the water source and is best determined by chemical or physicochemical tests. To vary the pH, the amount of soda ash and lime fed to the system can be changed.

The amount of hypochlorite used should be adjusted to yield a

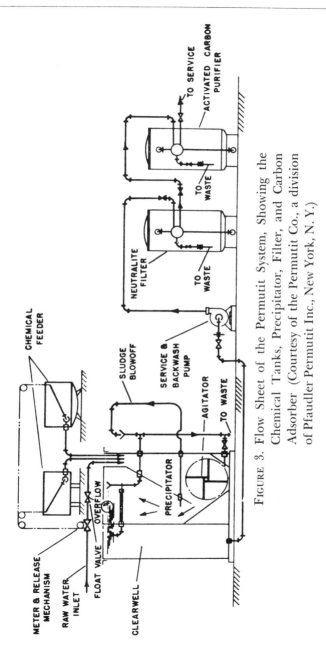

FIGURE 3. Flow Sheet of the Permutit System, Showing the Chemical Tanks, Precipitator, Filter, and Carbon Adsorber (Courtesy of the Permutit Co., a division of Pfaudler Permutit Inc., New York, N.Y.)

residual chlorine content of 8 to 10 ppm at the outlet of the calcite filter. If the amount of hypochlorite fed to the system is changed, several hours must be permitted to elapse for the system to be stabilized before making a test.

It may be pointed out here that the iron-coagulation scheme has one disadvantage in comparison to the alum-coagulation scheme and that is that if any iron, as mentioned, is left in the water from which the beverage is made, there will be an off-taste. It is essential to add hypochlorite to the coagulation system in the copperas scheme because the hypochlorite is required to oxidize the iron to the ferric state so that it can precipitate as ferric hydroxide.

In systems in which chemical coagulants are used, two visual tests can be made, if trouble arises, to check if the chemicals are leaking past the coagulation tank. Some of the chemicals will deposit on the activated carbon. If such a deposit is reddish in color, it is an indication that not all of the iron is being precipitated. If a white deposit is formed on the carbon, it is an indication that either alum or lime or both have passed the sand or calcite filters.

In coagulation systems, incomplete reactions may be due to a number of causes. A principal and increasingly serious cause is the presence of coagulation inhibitors, such as sequestering agents, in the water. Some municipalities add polyphosphates to their water supply and these inhibit the coagulation reactions. Synthetic detergents, sequestering agents of the chelate type, may also cause such inhibition. The need for proper adjustment of pH has already been mentioned. Other factors are the presence of minerals and proper temperature.

(b) Precipitator Stirrer

The precipitator stirrer is provided with speed regulators and adjustments. This is important for if the agitator speed is too low, the incoming water will not be mixed adequately with the chemicals and the improper results already discussed may occur. If the speed is too high, the flocs will be too fine and again some of the improper results mentioned will occur.

(c) Sludge Blanket

The sludge blanket should have a "concentration" of 25 per cent by volume. Concentration of sludge is defined as the settled volume of sludge in milliliters in a 100-milliliter cylinder. The amount of sludge in various parts of the precipitator is determined by filling a 100-milliliter graduated cylinder to the mark with the sample as it is drawn from a sampling pot or outlet. The sample is permitted to stand until the sludge settles and the volume of settled sludge is expressed as a percentage. Thus if a 100-milliliter sample had a settled sludge volume of 4 milliliters, the sludge concentration is 4 per cent by volume.

(d) Filters

The calcite or other filters should be backwashed in rotation at a time when there is no demand for treated water. The clear-water zone of the Permutit system will provide, generally, as mentioned, sufficient reserve water for this purpose. It may be necessary, however, to wait between washes for the clear water zone to refill in order to obtain sufficient water for a succeeding wash.

An adequate storage tank must be provided for storage of water if it is not practicable to plan the backwashing of filters when there is no demand for water. In such a system, a valve must be placed in the plant effluent line ahead of the storage tank and this valve must be closed off prior to backwashing.

The calcite and equivalent filters should be washed about once a week. If, however, there is a noticeable decrease in the rate of flow through a filter, caused very likely by clogging, the filter should be backwashed immediately. If such clogging occurs frequently, it means that the filter is not being washed often enough and the intervals between washings must be reduced.

When it is necessary to drain the filters to prevent freezing or for other reasons, the multiport valve is placed in the filter position, the inlet and outlet stop valves are closed, the waste valve is opened, and the drain plug is removed. Refilling must be done very slowly in order to avoid disturbing the gravel and calcite beds.

(e) Adsorbers

The activated-carbon adsorber or adsorbers should be washed about once a week. If, however, there is a noticeable decrease in the rate of flow through the adsorber, caused by clogging, the activated-carbon adsorber or purifier, as such devices are often called, should be backwashed immediately. If hypochlorite is used as one of the treatment chemicals, the adsorber effluent should be tested regularly. The outlet water should contain less than 0.1 ppm of available chlorine. If the chlorine content is greater than this value, the adsorber should be backwashed immediately. If either clogging or high available chlorine content in the treated water occurs frequently, the interval between backwashings must be reduced.

(f) Notes

It must be stressed that the precipitator and dealkalizer units are almost completely manual in operation. While the flow rate is regulated by the float valve in the inlet line, this condition holds only for those rates below the maximum, at which rate, the control valve is set. All of the other operations are completely manual. They are, fortunately, simple. Thus the filter multiport valve, by means of which the filter operations during a filter cycle are controlled, is a handwheel instead of the old-time nest of gate valves.

Despite the simplicity, the operation and care of one of these precipitator units requires, generally, the better part of one man's working day.

Mention has already been made, in the section on operation, of the need for care in the coagulation step and the need for checking to see that the chemicals do not leak past the system into the final water supply.

Water-purification systems such as the Permutit system can be adjusted to meet the specifications set by large carbonated-beverage bottling firms who franchise smaller dealers. Some bottling firms end their water-purification treatment by passing the water from the adsorber through a final polishing filter. Such a step makes certain that all suspended particles will be removed from the water.

b. **Electrolytic Coagulation**

In electrolytic-coagulation systems of water purification a gelatinous aluminum hydroxide is generated electrolytically. The "Electro-Chem" or "Oxylator" system which uses this method comprises a electrical coagulator, a sand filter, a dealkalizer, and an activated-carbon adsorber.

In this system, no chemicals are used in the water-treatment process. The coagulant, gelatinous aluminum hydroxide, is generated electrolytically from a battery of aluminum plates. This gelatinous

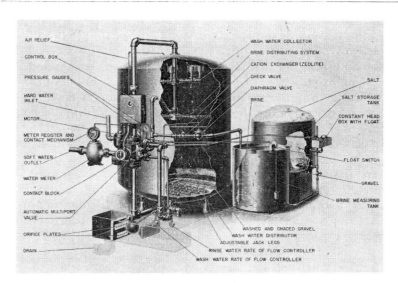

FIGURE 4. Permutit Fully Automatic Zeo-Karb Cation Exchanger (Courtesy of the Permutit Co., a division of Pfaudler Permutit Inc., New York, N. Y.)

coagulant adsorbs the colloidal matter, bacteria, algae, protozoa, plankton, and other types of microorganisms, and adsorbable organic matter from the water.

The treated water is then passed to a sand filter which contains seven layers of graded gravel and sand, with the top layer being a very fine sand. The filter is constructed in such a manner that the sand cannot be blown out during the backwashing step.

If it is necessary to reduce the alkalinity of the water, the treated water is sent from the sand filter to an acid-cycle ion exchanger, and finally, with or without the dealkalizer, to an activated-carbon adsorber. This adsorber contains six graded layers of sand and gravel with a top layer, 24 inches in height, of activated carbon.

In the electrolytic coagulant system, the gelatinous aluminum hydroxide is generated automatically in response to the water flow. No hypochlorite or other source of available chlorine is used for sterilization, because the microorganisms are adsorbed by the coagulant and filtered out of the water. Since no sterilizing agent is used to kill the microorganisms during the original treatment, it is necessary to use a "sanitizing" solution in connection with the backwashing treatments. This arrangement connects with a water-injection unit beneath the battery case of the electrolytic coagulator for delivery of the "sanitizing" solution to the sand filter and the carbon adsorber. This injection device can also be used to send the "sanitizing" solution to the bottling equipment and beyond the filters, through the carbonator, and filling machine, if desired.

Since no chemicals are used in this system, there is no danger of leakage of chemicals into the final beverage.

An elapsed-time meter is required to measure the length of time the system has been in operation. After about 50 hours the filters are backwashed to free them of the coagulant and its adsorbed impurities. At this time, the battery is reactivated by the addition of a special solution and the time of this treatment is controlled by a manual timer. After about 1500 hours of operation, which can be determined from the reading of the elapsed-time meter, it is necessary to replace the entire battery as a unit.

The sand filter and the carbon adsorber are backwashed to free them of the collected impurities and then they are washed with the "sanitizing" solution mentioned before, after which they must be left filled with treated water.

The electrochemical generation of gelatinous aluminum hydroxide is applicable to raw waters whose pH is in the range of 5 to 9. Such generation is independent of the temperature of the water, the mineral content of the raw water, and the presence of chemical coagulant inhibitors, such as the polyphosphates or synthetic detergents.

c. Cation Exchange

When a source of water is available which is suitable for bottled beverage manufacture, except that its alkalinity is too high, there is no need for a coagulation treatment or lime treatment. In such instances, an ion exchanger of the cation-exchange type in a hydrogen- or acid-exchange cycle may be used. Such treatment will not only diminish the alkalinity but will also decrease the hardness, for the following reactions take place:

$$H_2Z + 2Na^+ \underset{\text{regeneration}}{\overset{\text{reaction}}{\rightleftarrows}} Na_2Z + 2H^+$$

and

$$H_2Z + Ca^{++} \underset{\text{regeneration}}{\overset{\text{reaction}}{\rightleftarrows}} CaZ + 2H^+$$

in which H_2Z represents the cation exchanger in an acid- or hydrogen-exchange cycle.

In general this equipment, as shown in Figure 4, comprises a closed tank, containing the ion-exchange resin and the necessary supporting material, with the piping and valves necessary for forward flow and backwashing, but in addition lines must be provided for supplying the acid solution, kept in another tank or crock, for regeneration of the ion-exchange resin when it is exhausted.

The cation exchanger is backwashed once or twice a week, after which it is washed with the acid solution for regeneration. There is no need for the acid to be handled or measured directly by any operator, for special dispensing devices are available by means of which the acid can be added to the regeneration solution tank. Some of these tanks are provided with guide lines. The water is added to the lower of the guide lines and then the acid is pumped in to the second or higher line after which the solution is mixed. After regeneration the ion exchanger must be thoroughly rinsed with treated water. In some systems, the regenerative portion of the cycle requires only about 30 minutes. Such cation exchangers have a long life.

Since the hardness is reduced at the same time and may be reduced to zero hardness, the water to be used for bottling can be adjusted to any degree of alkalinity and hardness by blending the treated water with the untreated water, by means of a blending valve. Tests of the alkalinity of the water source must be made and then of the blended water and by this means the alkalinity of the final water can be maintained at a fairly constant level. From time to time, the blending valve must be reset, depending on the alkalinity of the raw water.

After reduction in alkalinity, the water should be passed through an activated-carbon adsorber to remove any residual taste and odor.

While soft water is not needed for the beverage water, it is very useful for bottle washing. This will be discussed in Chapter 13.

d. Deaeration

The treated water coming from the final filters contains air. As will be discussed in greater detail in Chapter 14 on spoilage and in Chapter 10 on carbon dioxide and carbonation, the presence of air in water has several disadvantages. Among these are that it reduces the ability of the water to retain carbon dioxide and it has a tendency to cause foaming in the filling step.

One method of deaeration is to place the water under a vacuum. One of these types of equipment [5] consists of a ¼-inch steel tank, coated with Heresite, a material which is sprayed on and baked to the shell to give it an enamel finish. A dome is bolted to the tank so that it can be removed for cleaning purposes. Trays containing 1 inch × 1 inch porcelain rings are placed in the upper portion of the tank. These rings break up the water and provide surfaces for the air to escape readily. The vacuum is produced by an oil-seal vacuum pump with a ¾ HP electric motor mounted on the base of the equipment (see Figure 5).

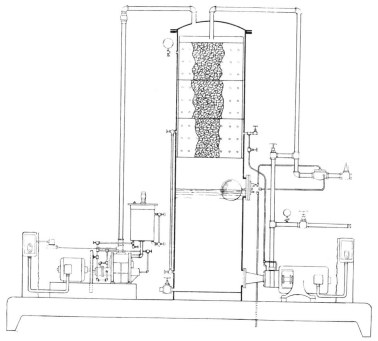

FIGURE 5. Syn-Cro-Mix Deaerator (Courtesy of the Geo. J. Meyer Manufacturing Co., Milwaukee, Wis.)

The filtered water is pumped into the dome at about 15 pounds of pressure by a pump of the monoblock type driven by a 1 HP motor, and is sent over the porcelain rings through a spray head, vacuum of 27 to 28 inches being maintained in the tank. The air released by the water is vented to the atmosphere, usually outside the building or into a portion of the plant away from the bottling area.

The deaerated water collects in a reservoir at the bottom of the tank where it is still under vacuum and from which it can be pumped to the proportionating equipment (See Chapter 11 on bottling and canning) under 15 pounds pressure. A float control actuates the inlet valve. An inlet is provided so that a chlorine or other sanitizing solution can be pumped through the deaeration tank to clean it.

A deaerator of this type can be placed in the basement under the bottling section or in other parts of the plant as may be most convenient.

LITERATURE CITED

1. *Public Health Service Drinking Water Standards 1946, U.S. Pub. Health Repts.* **61,** 371 (1946).
2. "Good Water for Good Beverages," New York, The Permutit Co., 1956.
3. "Oxylator New Water Treatment," Washington, Liquids Process Co., 1959.
4. Morris B. Jacobs, *Chemistry and Technology of Food and Food Products,* Vol. 1, Chapter II, 2nd Ed., New York, Interscience Publishers, 1951.
5. "Meyer Dumore Syn-Cro-Mix Beverage Filling System," Bull. No. SY. 126-2, Milwaukee, Geo. J. Meyer Manufacturing Co., 1952.

SELECTED REFERENCES

American Bottlers of Carbonated Beverages, *Plant Operation Manual,* Washington, 1940.

Standard Methods for the Examination of Water, Sewage, and Industrial Wastes. New York, Am. Pub. Health Assoc., 1955.

F. C. Nachod and E. Nordell, in M. B. Jacobs, Ed., *Chemistry and Technology of Food and Food Products,* Vol. III, Chapter LII, 2nd Ed., New York, Interscience Publishers, 1951.

H. E. Korab, *"Technical Problems of Bottled Carbonated Beverage Manufacture,"* Washington, Am. Bottlers of Carbonated Beverages, 1950.

H. B. Sliger, "Water Treatment in the Modern Bottling Plant," *Beverages,* (April 1953).

I. M. Markwood, "Good Beverages Need Pure Water," *Natl. Bottlers' Gaz.* (February 1953).

Chapter 6

FLAVORS AND FLAVORING

1. FLAVOR SENSATION

The flavor sensation is in reality a complex of sensations. Thus, for example, the flavor of a ginger ale is the resultant of the sensations attributable to the sweet taste of the sugar or other sweetening agent, the sour taste of the acidulant, the pungent taste of the ginger, the taste and odor sensations aroused by the other flavoring components, the tickling sensation caused by the impingement of the bubbles of escaping carbon dioxide gas against the walls and roof of the mouth and the tongue, the sensation of coldness due to the low temperature of the beverage, the body effect of the sugar, and finally the feeling of satisfaction given by a good soft drink.

It can be seen, then, that the flavor of a carbonated beverage is the sum of the flavoring action of the various components of the soft drink, namely, the sugar, the acid, the carbon dioxide, and the flavoring materials themselves. Although in previous chapters the significance of sugar, sweetening agents, and acid in the preparation of beverages was pointed out, this will bear some slight repetition.

a. **Sugar**

As mentioned the sugars and sirups used in the preparation of soft drinks play a part in the flavor sensation, but their actual role is more than that of mere sweetening agents. In addition to the sweet taste, sugars have three other principal flavor functions, namely, (a) sugars provide the sweetness necessary to balance the acid and other taste-producing components properly and thus produce a balanced flavored soft drink; (b) sugars furnish sufficient body to raise the beverage out of the sweetened watery class; and (c) sugars serve to carry the flavor and thus deposit it uniformly when consumed.

b. Artificial Sweeteners

Artificial sweeteners, like saccharin and cyclamate, give the sweet taste required in "dietetic" soda pop without adding food value, but they do not assist in giving the beverage a "body" effect. To obtain this body effect in artificially sweetened carbonated beverages, bodying agents like sorbitol and carboxymethylcellulose are used. See Chapter 8.

c. Acid

The amount and kind of acid used in a carbonated beverage often, as mentioned in the chapter in which acids used in soft drinks were discussed, have a marked influence on the flavor of the beverage. This is due not only to the pH of the drink but also to the character of the acidulant. Thus, citric acid is the acid that is characteristic of the citrus fruits. Tartaric acid is the acid characteristic of grapes. Using such acid in the appropriate fruit drink is advantageous in producing the flavor desired.

d. Flavors

While both the sweetening agent and the acid do influence markedly the flavor of a given soft drink, the characteristic flavor results from the flavoring materials added. The principal flavors used by the beverage manufacturer come in the form of alcoholic extracts or essences, aqueous solutions and emulsions, solutions of flavors in glycerol and propylene glycol, and fruit-juice concentrates. Soft drink flavor ingredients of an essential-oil character often require extraction with alcohol from the source raw material. Water-soluble flavoring components can be put up as aqueous solutions. When necessary because the water solubility is low, emulsions can be prepared or solutions in glycerol or propylene glycol can be made. Such emulsions or solutions avoid the use of alcohol. Concentrated fruit juices are used in preference to fruit juices themselves, for on dilution with the carbonated water greater fidelity of fruit flavor can be obtained.

It is necessary to point out that for beverages which have country-wide distribution, the complete flavored sirup is often prepared at central plants and is distributed from such central locations to the bottling plant.

2. CLASSIFICATION OF FLAVORING MATERIALS

Flavoring materials can be classified in several ways. Possibly the simplest classification is to place the flavors into three major groups, namely, natural flavors, fortified flavors, and synthetic or artificial flavors and, of course, flavoring materials. Such a classification is not adequate for the purposes of this book for a more detailed classification will provide more useful information.

a. Naturally Occurring Flavoring Materials

First, there are naturally occurring flavoring substances of an organic nature, such as spices and herbs. These are, in general, derived from flowers, leaves, buds, fruits, roots, tubers, bark, stems, and wood of various plants. Few of such materials are used for the flavoring of soft drinks, but a number of the products derived from this group are employed.

b. Inorganic Natural Flavoring Materials

Second, there are the naturally occurring flavoring materials of an inorganic nature. The principal one of these insofar as carbonated beverage manufacture is concerned is, of course, common salt, but in the manufacture of club soda and mineral waters a number of other salts are also employed.

c. Natural Flavors

Third, there is a large group of flavoring materials which are separated from natural products. These can be placed into two major subgroups. On the one hand is the group of essential oils and oleoresins used for flavoring and on the other are products, such as alcoholic flavor essences and extracts, and concentrated natural fruit flavors.

d. Isolates

Fourth, there is the group of substances isolated from naturally occurring materials. Such substances are known as *isolates* and while they are still important, their relative importance has diminished. Representative substances in this group are anethole, cineole, eugenol, cinnamaldehyde, and citral. These are obtained respectively from anise and star-anise oils, eucalyptus oil, clove oil, cassia and cinnamon oils, and from lemon and lemongrass oils. Menthol from peppermint oil is another example.

e. Semisynthetics

Fifth, there is the group of flavoring substances known as semisynthetics. These have been synthesized using isolates as the starting materials. They, too, were a most important group, but their relative importance is also diminishing. Representative members of this group are isoeugenol from eugenol, vanillin from eugenol and safrole, piperonal from safrole, zingerone from gingerol, the pungent component of ginger, and caffeine from theobromine, the characteristic alkaloid of cacao.

f. Synthetics

Sixth, there is the most important group of synthetic flavoring substances. Actually this group is best divided into two major subgroups. The first of these comprises those synthetic substances made from coal, petroleum, water, and air or from simple organic substances made from these raw materials, which are identical, chemically speaking, with flavoring substances isolated from natural sources. The one of greatest historical interest is vanillin. Others that fall into this group are cinnamaldehyde, biacetyl (diacetyl), and methyl beta-methylthiolpropionate, a flavoring substance present in pineapple.

The second subgroup of synthetic flavors that is certainly of great importance to the beverage manufacturer comprises a host of flavoring substances which have no counterpart in nature. Generally, these are extremely powerful flavors having many times the flavoring power of the natural products they resemble. It is a common misconception that such products are "new" in the sense that they are the result of chemical research done within the past two decades. The fact is that many of these substances were prepared and even used as flavoring substances some fifty years ago. Representative of this group are gamma-undecalactone that is so-called Aldehyde C_{14} or peach aldehyde, ethyl methylphenylglycidate or ethyl phenylglycidate that is ethyl alpha,beta-epoxy-beta-methylhydrocinnamate or ethyl alpha,beta-epoxyhydrocinnamate, the so-called Aldehyde C_{16} or coconut aldehyde, isobutyl furylpropionate (reminiscent of pineapple) and a host of other substances. Many of these will be noted in the specific formulations in the text.

3. CRITERIA FOR BEVERAGE FLAVORS

There are certain criteria that a beverage flavor, whether prepared by a flavor manufacturer or by the beverage bottler, must meet. These are solubility, flavor fidelity, resistance to acidity, noncontamination, and certain other criteria of lesser importance.

a. Solubility

Possibly the most important criterion for a flavor for beverage use is that of solubility. Since this is an essential characteristic, it is often necessary to eliminate certain relatively water insoluble components normally present in a natural flavoring material like an essential oil, before it can be used for a particular beverage. For instance, if a perfectly clear lemon- or orange-flavored carbonated soft drink is desired, it is necessary to use citrus flavors in which the terpene concentration has been reduced. Terpeneless and sesquiterpeneless flavoring oils are in this category.

For cloudy type orange or other citrus beverages, the elimination of terpenes and other insoluble flavor components, is, of course, far less important.

An additional factor to consider is that, if the flavoring is not completely soluble and a perfectly clear carbonated beverage is desired, it becomes necessary to filter the sirup or even the beverage. This not only adds an otherwise unnecessary step to the bottling process but actually results in a loss of flavoring material.

b. Flavor Fidelity

Since carbonated nonalcoholic beverages and other beverages are in contact with the mouth for only a transitory period of time, the need for flavor fidelity is not great, certainly not as great as in the case of a product which stays in the mouth for some time as, e.g., candy does.

Indeed in the case of soft drinks or soda pop, one often does not desire a lingering taste, even a lingering taste of high fidelity. Therefore, if such a taste were present it would be a decided disadvantage to the beverage. The lack of need of flavor fidelity in carbonated beverage flavors was exploited to such an extent years ago that the flavor of some beverages, notably raspberry, had virtually no resemblance to the natural flavor. This "artificial" taste was so widely accepted that soft drinks flavored with true raspberry fruit flavor were not popular.

In more recent years, considerable progress has been made in this respect. Natural flavors have a much greater vogue and the use of "true" fruit flavors often gives a competitive advantage.

c. **Acid Resistance**

Since many beverages have relatively low pH, that is, relatively high acidity, it is necessary that the flavoring components are resistant to decomposition attributable to the acid or to accelerated oxidation because of the acid.

d. **Noncontamination**

It is most important that the flavoring of a soft drink does not contaminate the beverage because, as noted in other sections, such drinks are generally not heat treated. When alcoholic extracts are employed, particularly if the concentration of the alcohol is over 20 per cent, the alcohol will exert a preservative action. In some flavors the acid concentration will be sufficiently high to reduce the danger of contamination. In the case of those beverages which are flavored with fruit concentrates that contain fruit pulp and other insoluble matter, the problem becomes more complex and in some instances a chemical preservative, like sodium benzoate, must be used.

e. **Other Criteria**

Resistance to destruction by heat or thermal hazard is generally not an essential characteristic for a beverage flavor, as seldom are soft drinks heat treated and thus the flavors used are not subject to this type of decomposition. Beverage flavors must, however, be able to withstand temperatures of the order of 100°F in an acid environment.

The possibility that a flavor will be inadequate because of dilution is also not too important for usually carbonated beverage flavors are made sufficiently concentrated so that sufficient character is retained after dilution. This, however, is not true if unconcentrated fruit juices are used, for in such instances too great dilution may diminish the flavor strength beyond an acceptable degree. To counteract such dilution, it is common to add the characteristic essential oil, such as orange oil to an orange juice concentrate flavor, etc.

The presence or absence of alcohol in the flavor has its greatest effect on the solubility of the flavoring materials themselves, in the alcoholic prepared flavor, in the flavored sirup, and even in the final beverage. If alcohol is used, the flavor will probably be more soluble in all of the products mentioned.

4. NATURAL FLAVORINGS

The natural flavors used by the flavor industry can be considered in several categories principally as (a) essential oils, (b) oleoresins, (c) flavor extracts, (d) flavor essences, and (e) fruit flavors. In a sense, the first two can be placed in one group and the last three can be combined into another group as true fruit flavors.

a. Essential Oils

Essential oils are the products obtained from the plant kingdom in which the odoriferous and flavoring characteristics are concentrated. There are four principal groups of essential oils with respect to the method of preparation. These are first, oils obtained by distillation; second, oils derived by solvent extraction; third, oils separated by expression; and fourth, those prepared by enfleurage. For use in beverages, the more important methods are distillation (ginger oil, sweet birch oil, wintergreen oil, and some citrus oils, particularly lime oil, are obtained by distillation) and expression (the citrus oils orange, lemon, tangerine, lime, and grapefruit are made by pressing the rinds of the fruit).

It was mentioned in a previous section of this chapter that the solubility of a flavoring was most important. Guenther [1] points out that most essential oils are mixtures of hydrocarbons, such as terpenes and sesquiterpenes; oxygenated compounds, such as esters, alcohols, ethers, aldehydes, ketones, lactones, phenols, phenol ethers, and compounds of more than one functional group; and a small amount of relatively nonvolatile substances, such as waxes, paraffins, and like materials.

The oxygenated compounds are the main flavoring components,

although the terpenes and sesquiterpenes also have some flavor value. The oxygenated compounds are more soluble in water than the terpenes and sesquiterpenes, which are hydrocarbons of the general formula $(C_5H_8)_n$ where n is 2 for the terpenes and 3 for the sesquiterpenes; thus if essential oils, such as the straight citrus oils, are used there is the chance that the terpenes and sesquiterpenes will be precipitated from solution. This will be undesirable in a clear type beverage. To avoid such difficulties, terpeneless and sesquiterpeneless oils are used for flavoring. Terpeneless oils are essential oils from which the terpenes have been removed and sesquiterpeneless oils are essential oils from which both the terpenes and the sesquiterpenes have been removed. The removal can be accomplished by special distillation *in vacuo* or by solvent-extraction methods. Since the oxygenated flavor components have been concentrated in terpeneless and sesquiterpeneless oils, they are more powerful flavors than the original oils. Examples of the use of essential oils are given in Chapter 8 on emulsion flavors.

b. **Oleoresins**

Oleoresins of the pharmaceutical type are solvent extracts of a given plant from which the major portion of the solvent is removed by distillation and the remainder of the solvent by evaporation, leaving a dark, soft mass. The principal oleoresins used for the flavoring of beverages are oleoresin ginger and oleoresin capsicum.

c. **True Fruit Flavors**

True fruit flavors can be made by alcoholic distillation, flash vaporization and distillation,[2] maceration, and percolation of the fruit and by concentration of the crushed fruit and fruit juices. It is beyond the scope of this book to discuss these methods in detail for the average beverage manufacturer usually purchases these products for flavoring his beverages rather than makes them. A distinction is made between the terms flavor essence, flavoring extract, and flavor.

A flavor essence may be defined as a solution of a flavor ester mixture or of flavoring materials and substances in alcohol. Because of the wide use of other solvents and diluents, the term flavor essence is often considered to be a solution of a flavor ester mixture or other flavoring material in an appropriate solvent.

The Food and Drug Administration distinguishes between the

terms *flavoring extract* and *flavor*. The vehicle, solvent, or menstruum of a flavoring extract is ethyl alcohol of proper strength. The terms "extract" and "flavor" are not synonymous for the first term implies an alcoholic solution or product. Flavoring products prepared with solvents or vehicles other than ethyl alcohol should be labeled with the term "flavor" but even when this term is used, the resulting product should contain the same kinds and proportions of flavoring ingredients as are contained in the product prepared with alcohol. Thus a lemon flavor should contain the same amount of lemon flavoring oil as lemon extract.

Genuine concentrated fruit extracts must be made from the extractives of the fruit in a nonflavoring medium, such as ethyl alcohol or water, and must conform to the fruit named on the label. Usually they are depectinized and other colloidal materials are removed while the manufacturing process is carried on in such a manner as to retain the maximum amount of color. Such products are freely soluble in water and in beverage sirups.

Depending on the fruit flavor, 6 to 8 ounces per gallon of sirup is recommended as the flavoring ratio. Among the true fruit flavors commonly used are blackberry, cherry, grape, raspberry, and strawberry.

A second type of true fruit flavor is available for bottlers. This is generally labeled "fruit flavor (name of specific fruit) with other natural flavors." Such products contain at least 50 per cent of their flavor strength from extractives of one fruit source, the one named, and not more than 50 per cent of other natural flavors. The addition of other natural flavorings enables the flavoring strength to be increased.

Usually 3 to 4 ounces of such flavors are used per gallon of 31° to 32°Bé sirup with 1.5 to 2.0 ounces of 50 per cent citric acid solution.

Blangsted [3] quotes Katz (1952) who describes the preparation of true fruit flavors for use in the manufacture of beverages as follows:

"The fruit is crushed and put through a hydraulic press to recover the juice. The dry press cake retains some of the volatile flavors, and these may be extracted by steam distillation, to be added later to the concentrated juice. Concentrates prepared for use in the beverage industry must dilute sparkling clear and it is, therefore, important to depectinize the freshly pressed juice before starting the concentration process. Enzymatic reaction on the juice by the addition of such enzymes as pectase and clarase effect complete clarification within 4 days, when carried out under careful control of pH and temperatures. The juice is filtered through hydraulic filter presses equipped with pumps to remove all the sediment formed during the clarification process. The juice is concentrated under reduced pressure at temperatures between 95° and 104°F. To recover the volatile fractions, the equipment is provided with a condenser and receiver kept cool by refrigeration. The juice passes through a heat exchanger at great velocity before entering the evaporation chamber, to assure that there will be no loss of color during concentration. After the volatile aromatics are distilled off, the juice is concentrated to the desired strength. The condensed distillate containing the aromatics is added to the finished concentrate which then possesses the true flavor of the fresh fruit. To strengthen the flavor of these concentrates, a mixture of other natural flavors may be added, which is compounded from essential oils and extractives of herbs, such as Saint John's bread and vanilla bean. It usually requires 4 ounces of a true fruit concentrate with other natural flavors to flavor 1 gallon of bottling sirup. Grapes, raspberries, blackberries, cherries, and other fruits may be processed in this manner."

A true fruit raspberry flavor extract [4] for carbonated beverages can be made on a small scale as follows: Place 4 pounds of dried raspberries into a cylindrical percolator. Cover the fruit with 25 per cent alcohol and allow the mixture to macerate for a day or two. Start the percolation by withdrawing the percolate very slowly, maintaining the level of the menstruum in the percolator above the fruit with 25 per cent alcohol. Continue the percolation for several days, alternating percolation and maceration, but making certain to maintain the level of the menstruum above the fruit at all times. Continue the process until 1 gallon of percolate has been obtained. Since there is still some flavor and color remaining in the fruit after the first completed percolation, it is well to repeat the percolation with 25 per cent alcohol until a second gallon of percolate is obtained. The second percolate can be used to prepare the flavor from a new batch of dried raspberries.

To prepare the bottling concentrate for use with non-nutritive sweeteners, mix 6 fluid ounces of the true fruit raspberry extract made as detailed in the preceding paragraph with 2 fluid ounces of 50 per cent citric acid, 1.0 to 1.5 fluid ounces of a red color solution, and 2.5 ounces of a cyclamate, such as Sucaryl calcium, in sufficient water to make a gallon of concentrate. A throw of 1 ounce of this bottling concentrate to a 7-ounce bottle with carbonated water containing 2.0 to 2.5 volumes of carbon dioxide will yield a beverage with 0.27 per cent of calcium cyclamate.

This true fruit raspberry flavor can be fortified with other natural flavors [4] in the following manner: Prepare a tincture of other natural flavors by placing 1.5 ounces of dried raspberries, 0.75 ounce of granulated rhatany root, and 0.75 ounce of granulated orris root into a separatory funnel and cover with 32 fluid ounces of 75 per cent alcohol. Allow to macerate for 2 or 3 days and draw off the liquid. Add sufficient 75 per cent alcohol to this extract to make 1 quart. To this quart of tincture add 4 fluid ounces of alcoholic orris solution prepared by dissolving 0.25 ounce of orris concrete in 8 fluid ounces of alcohol, 2.5 milliliters of clove oil, 0.4 milliliter of oil of green cognac, 3.6 milliliter of cinnamon oil and 1 fluid ounce of five-fold lemon oil. Mix thoroughly.

Add 2 fluid ounces of the prepared tincture of other natural flavorings to 126 fluid ounces of true fruit raspberry flavor to make 1 gallon of the true fruit raspberry flavor with other natural flavors.

The bottling concentrate and the non-nutritive sweetened carbonated beverage are made in a manner analogous to that detailed.

5. FORTIFIED FLAVORS

True fruit flavors do not have, at times, the flavoring power desired. Some attempt has been made to increase flavor strength by the addition of other natural flavors as has been mentioned in the preceding section.

Intensification or fortification to a greater degree can be obtained by the addition of powerful flavoring compounds, for instance, beta-ionone in the case of raspberry flavor and allyl caproate in the case of pineapple flavor.

In the manufacture of fortified flavors, sometimes all of the flavor extractives and the sugars are retained, but in other preparations the sugars are removed. It is customary to employ fortified flavors in the ratio of 1 to 2 ounces of flavor per gallon of sirup.

6. ARTIFICIAL FLAVORS

The recognition of the similarity of the flavor and odor of organic compounds synthesized and isolated during the first flush and growth of organic chemistry in the nineteenth century, to that of natural flavoring substances led to the use of alcoholic solutions of these isolated and synthesized substances as artificial flavors. In the latter part of the nineteenth century and in the early part of the twentieth century, as has been discussed by Jacobs,[5] substances were prepared synthetically which were not found in any natural source and which had many times the flavoring power of the analogous natural flavor types.

It should be stressed that the use of artificial and imitation flavors was so common about 1900 in the beverage industry that, with the exception of lemon, orange, lime, and the complex mixtures of the root beers group, the ginger ales, and the kolas, nearly all the flavors were made with synthetic flavors. Indeed it has been pointed out that artificial raspberry flavored carbonated beverage was so widely accepted that consumers were unable to recognize the true fruit beverage and considerable effort had to be expended to get the latter accepted.

Flavoring substances are used to prepare certain basic flavor mixtures which may be termed, for the purposes of identification and convenience, *flavor ester mixtures*. It is to be understood that not all the components of such mixtures are esters, but the term seems preferable to other terms, such as the "basic ethers" of Walter,[6] the "fruit basis" of Gazan,[7] and the "fruit oils," a designation in common use by commercial establishments.

Flavor ester mixtures are then mixtures of flavoring materials and substances which closely resemble the odor and flavor of natural oils obtained from fruits, seeds, nuts, roots, stems, leaves, herbs, spices, etc., and that contain no solvent or diluent like alcohol, glycerol, or propylene glycol.

a. **Flavor Components**

In section 2 of this chapter, the various materials used for flavoring were classified. The flavor components used for the preparation of flavors for beverages can most conveniently be classified according to the common functional or chemical groups, that is as esters, aldehydes, ketones, ethers, lactones, alcohols, acids, anhydrides, hydrocarbons, terpenes, and mercaptans. These compounds may belong either to the aliphatic or aromatic chemical groups and they may be either saturated or unsaturated.

The mere mention of these various groups and a glance at some of the formulations given in this text will make it clear to the beverage manufacturer that considerable thought must be given to the preparation of such flavors in the plant. There are unquestionable advantages in the preparation of one's own flavor formulations, such as gains in economy, secrecy, development of an independent line, etc., but for uniformity of supply and convenience, it may be better to purchase flavors from a supplier.

b. **Compounding**

It is very likely true that the compounding of flavors and flavor ester mixtures is as much an art as it is a science. According to some connoisseurs, the addition of a single excess drop of a strong flavoring material may spoil a flavor composition just as a single excess drop of a perfume component may ruin a perfume formulation. This opinion is probably somewhat exaggerated, but it does serve to emphasize the need for care and accuracy in the preparation of a flavor mixture.

Persons who have had little experience in the compounding of flavor formulations are prone to underestimate the strength of many synthetic flavoring substances and essential oils. Consequently, there is a tendency to add too much of such substances to their formulations.

Another common error is that of failing to note the difference in aroma strength—that is, the aroma quality of certain raw materials, so that it becomes difficult to duplicate successful composition merely by using similar quantities of these raw materials. This used to be more of a problem than it is currently because of more complete control and standardization of raw materials by suppliers. It was also much more of a problem with natural than with synthetic raw materials since the former are more difficult to standardize to a uniform quality.

In order to overcome these difficulties, it is best to make small-scale experimental batches, noting the exact proportion of each component used. By variation of the formula, with such small-scale trials, the changes to be made in the preparation of the commercial-scale batch will become apparent. The ability to reproduce the same flavor effect from ingredients which differ in quality, as well as the formulation of entirely new flavor effects is undoubtedly an art. For the beverage manufacturer, in contrast to the flavor manufacturer, this is not too great a difficulty, since he will, in general, prepare only a limited number of flavors for his own use.

(1) Role of Components

Each flavor component of a flavor formulation has its role to play and a proper understanding of that role is a part of the science of compounding flavors. Natural flavors, as well as synthetic substances, are among the ingredients to be used.

In order to simulate the flavor of natural flavoring materials or to prepare other beverage flavors of acceptable quality, it is necessary to make an adequate blend of synthetic flavoring substances, for natural flavors are also blends of flavoring ingredients. In general, a flavor ester mixture has a group of major ester components. These give the flavor its characteristic aroma, such as fruit, like raspberry, or spice, like ginger, or other aroma, like kola. A small amount of a powerful flavoring substance, usually a lactone, hydroxycinnamate, aldehyde, or ketone, with a similar characteristic aroma is often added to intensify the flavor. For raspberry flavor, for example, beta-ionone is the intensifying ingredient. When these substances are added to natural flavors, as mentioned in a prior section, the flavors are said to be fortified. Modifying substances, such as aldehydes, ketones, ethers, compounds with more than one functional group, such as vanillin, and some esters are also added. Other components like terpene alcohols and the esters of these alcohols are added to give tone or nuance or to lend a flowery note to the mixture. Still other components, termed fixatives, are added to prevent uneven volatilization of the various ingredients. All of these, intensifiers, modifiers, toners, and fixatives, play a role in making the final blend really resemble the "true" fruit flavor or bring out the aroma desired.

(2) Blending

In the following chapter many examples of bottlers' flavor formulations will be given. Here just one example will be used to show the influence of each component in making the final blend.

RASPBERRY FLAVOR ESTER MIXTURE [a]

Flavor Component	Parts
Isobutyl acetate	425
Isoamyl acetate	275
Ethyl acetate	200
Ethyl formate	35
Benzyl benzoate	20
Bromelia	15
Vanillin	10
Linaloöl	10
Eugenol	6
Benzyl acetate	2
Geraniol	1
Ionone	1
Total	1000

Dilute with 2 volumes of alcohol to make the essence

[a] After Morris B. Jacobs, *Synthetic Food Adjuncts*, Van Nostrand, New York, 1947.

It will be noted that the principal components of the raspberry-flavor ester mixture are the esters isobutyl acetate, isoamyl acetate, ethyl acetate, and ethyl formate. Ionone serves as the intensifier; the modifiers are vanillin, bromelia, and eugenol; linaloöl, geraniol, and benzyl acetate are toners and lend the flowery note; and benzyl benzoate serves as the fixative. The formulations in the next chapter can be interpreted in a similar fashion.

LITERATURE CITED

1. E. Guenther, *The Essential Oils*, New York, Van Nostrand, 1948.
2. H. P. Milleville in D. K. Tressler and M. A. Joslyn, *The Chemistry and Technology of Fruit and Vegetable Juice Production*, New York, Avi Publishing, 1954.
3. S. Blangsted in D. K. Tressler and M. A. Joslyn, *The Chemistry and Technology of Fruit and Vegetable Juice Production*, New York, Avi Publishing, 1954.
4. *"Sucaryl" Sweetened Beverages*, Technical Information No. 42, North Chicago, Abbott Laboratories, 1957.

5. M. B. Jacobs, *Synthetic Food Adjuncts,* New York, Van Nostrand, 1947.
6. E. Walter, *Manual for the Essence Industry,* New York, Wiley, 1916.
7. M. Gazan, *Flavours and Essences,* New York, Van Nostrand, 1936.

SELECTED BIBLIOGRAPHY

M. B. Jacobs, ed., *Chemistry and Technology of Food and Food Products,* 2nd Ed., Interscience Publishers, New York, 1951.

S. Blumenthal, *Food Manufacturing,* New York, Chemical Publishing, 1942.

M. B. Jacobs, *Am. Perfumer Essent. Oil Rev.,* **55,** 41 (1950) ; **55,** 131, 215 (1950) ; **55,** 387, 489 (1950) ; **56,** 43, 309 (1950) ; **58,** 128 (1952).

Chapter 7

SPECIALTY AND FRUIT FLAVORS

In Chapter 1, it was noted that the carbonated beverages could be divided into three flavor groups; (1) the specialty-flavored group, such as the ginger ales, root beers, kolas, cream sodas, and the like; (2) the fruit-flavored group, including the citrus-flavored drinks; and (3) the mineral-flavored and nonflavored group, including club soda, plain seltzer, or carbonated water, etc.

In the preceding chapter, the technique of flavor formulation was discussed and in this chapter the formulation of some of the members of the foregoing groups will be considered.

1. SPECIALTY-FLAVORED GROUP

The importance of the specialty-flavored soft drinks has been discussed in Chapter 1 and the discussion need not be repeated here. They comprise the greater portion of all the carbonated beverages sold in the United States. The kolas and ginger ales have world-wide distribution.

a. Root Beer and Sarsaparilla

It is likely that root beer was prepared shortly after the introduction of carbonated soft drinks in the late eighteenth and early nineteenth centuries for root beer has certainly been known over a century. For instance, mention is made of this beverage by Nathaniel Hawthorne in his book, "The House of Seven Gables," which was published in 1851. Hawthorne writes, "No less than five persons, during the forenoon, inquired for ginger beer or root beer or any kind of a similar beverage and, obtaining nothing of the kind, went off in exceedingly bad humor."

Actually the root beer of yesteryear was different from the product we call root beer, at present. Before 1877, it was customary for the consumer to purchase a ground mixture of roots, herbs, and bark and steep the mixture in boiling water to prepare an extract. Sugar and yeast were added to the extract and the mixture was allowed to ferment with the subsequent production of alcohol and carbon dioxide to make an alcoholic beverage. About 1880, one processor prepared the extract himself and sold it for use by bottlers and in places where soda was dispensed.

About 1905, a prepared sirup containing the root, herb, and bark extract with additional flavoring ingredients to which only carbonated water had to be added to make a complete beverage was furnished by another processor. This preparation, however, contained no alcohol and thus this type of root beer was in reality a soft drink.

Root beer flavoring components may be placed into three principal categories, namely, (1) plant components, (2) essential oil ingredients, and (3) aromatic chemicals including both isolates and synthetic organic chemicals.[1]

The botanical components may consist of sarsaparilla, which is generally the most important ingredient and licorice, wintergreen, sassafras, hops, ginger, coriander, dandelion, root, wild cherry bark, althea, angelica, allspice, and yellow dock which are used in lesser quantities as modifiers and blenders. The pungent, acrid, bitter, and aromatic flavors of these ingredients are combined with the characteristic flavor of sarsaparilla to give the principal flavor combination known as root beer.

Actually, the essential oil components of root beer flavor are also of plant origin but the use of the plants or the plant parts themselves or extracts of the plants, on the one hand, and the essential oils derived from the plants, on the other hand, in the formulation of root beer flavors is a useful distinction because use of the first group is believed to give the final beverage a better flavor. The essential-oil ingredients may consist of wintergreen oil, sweet-birch oil, clove oil, nutmeg oil, and anise oil.

In addition to these groups, there is the third group of aromatic chemicals, such as methyl salicylate, ethyl salicylate, anethole, and eugenol that are almost always used as components of such formulations. It can be seen that each of these is a component of the essential oils mentioned in the preceding paragraph, but the organic compound actually used may be a synthetic one rather than an isolate.

Federal definitions for root beer flavor, birch beer flavor, and sarsaparilla flavor were promulgated under the old Wiley Pure Food Act of 1906, but these definitions were not reissued under the authority of the Food, Drug, and Cosmetic Act of 1938 (see Chapter 1). In these old definitions, it may be noted, that methyl salicylate or wintergreen oil, or sweet-birch oil, and sassafras oil were principal ingredients of the flavor. It is to be stressed here again that in 1957 attention was called to the relative toxicity of safrole and sassafras oil, of which safrole is a principal component, and related essential oils in which safrole is a major component as, for instance, Brazilian sassafras oil (oil of *Ocotea cymbarum*) and oil of *Illicium parviflorum*. It is probable that rules may be promulgated banning the use of safrole or safrole-containing essential oils in carbonated beverages or rigidly controlling such use.

The use of ingredients of the first two groups in a root-beer flavor formulation can be illustrated by the following composition.[2]

ROOT-BEER FLAVOR SIRUP

Fluid extract of sarsaparilla	4 pt
Fluid extract of licorice	4 pt
Fluid extract of ginger	2 pt
Fluid extract of althea	2 pt
Fluid extract of angelica	8 oz
Fluid extract of soap bark	8 oz
Tincture of vanilla	1 pt
Wintergreen oil	4 oz
Sassafras oil substitute	4 oz
Anise oil	2 oz
Orange oil	2 oz
Nutmeg oil	4 dr
Clove oil	4 dr
Caramel coloring	3 gal
Sirup	30 gal

This flavored sirup is generally used in the proportion that is, a throw of 1¼ ounces plus sufficient carbonated water to make an 8-ounce bottled beverage. This formulation makes use of fluid extracts of the plants characteristic of root beer flavor which, as mentioned, is considered an advantage.

A standard formulation, formerly in the U.S.P. but currently in the *National Formulary* [3] is the "Compound Syrup of Sarsaparilla." This is a relatively simple formulation that can be prepared quickly.

SARSAPARILLA SIRUP

	milliliters
Fluid extract of sarsaparilla	200.0
Fluid extract of glycyrrhiza	15.0
Sassafras oil substitute	0.2
Anise oil	0.2
Methyl salicylate	0.2
Alcohol	19.4
Sirup	765.0

The fluid extracts are mixed together and the oils are dissolved separately in the alcohol and then added to the extracts. The combined mixture is stirred until the oils are dissolved and then is added slowly to the sirup, with constant stirring.

Typical of the formulations in which only essential oils and aromatic chemicals are used in a root beer flavor are the following:

ROOT-BEER FLAVOR

Wintergreen oil	2.5 oz
Sassafras oil substitute	2.5 oz
Sweet orange oil	2.0 oz
Amyl butyrate	2.0 oz
Spruce oil	0.5 oz
Clove oil	2.0 dr
Anise oil	2.0 dr
Cologne spirits	7.0 pt
Water	2.0 pt

The oils are added in small portions to the Cologne spirits, the mixture is well stirred, after which the water is added. Then the mixture is filtered through pumice. This is a 1½ ounce flavor so that 1½ ounces is used to 1 gallon of sirup to make the flavored sirup.

ROOT-BEER FLAVOR

Wintergreen oil	20.0 oz
Sassafras oil substitute	24.0 oz
Anise oil	10.0 dr
Cassia oil	1.0 oz
Cologne spirits	3.5 gal
Water	1.0 gal

This formulation can be handled in the same manner as detailed after the preceding formula. It is a 2-ounce flavor and thus 2 ounces are used for each gallon of sirup.

Since root beer is not considered a member of acid-type beverages, as discussed in Chapter 4, it is generally not advisable to add citric acid. The pH is kept to about 5 and, if necessary, phosphoric acid may be used to obtain this pH. The last two formulations mentioned are not considered as good as those in which fluid extracts of plants are used.

b. **Kola Flavors**

The principal kola, "ola" or cola flavors are secrets. Most of the bottled kola drinks are distributed by local franchised bottlers who obtain the finished flavor or the flavored sirup from large beverage organizations and add it to their own sirup and carbonated water or carbonate directly. These flavor formulations are mixtures of a number of ingredients containing many components so that they are difficult to analyze and duplicate. It may well be that a small distributor can purchase the finished kola flavor at less cost than his cost of individual manufacture of the flavor would be, particularly if the labor involved in the flavor preparation is included in the cost.

A principal characteristic of kola beverages is that they contain caffeine. Many years ago, Dr. Harvey W. Wiley (1919), in his book on *Beverages and Their Adulteration*,[4] discussed soft drinks containing caffeine. In his discussion Wiley said:

> "The chief soft drink which contains caffein is coca cola. In addition to the content of caffein, coca cola also contains the extract from coca leaves from which the cocain has been previously obtained. In addition to this, also, a very small quantity of cola nut is employed, together with sugar, caramel and acid and aro-

matic substances. The composition of coca cola syrup as disclosed before the federal court in Chattanooga is as follows:

"Caffein (grains per fluid ounce)	0.92– 1.30
Phosphoric acid (H_3PO_4) (percent)	0.26– 0.30
Sugar, total (percent)	48.86–58.00
Alcohol (percent by volume)	0.90– 1.27
Caramel, glycerin, lime juice, essential oils, and plant extractives	Present
Water (percent)	34.00–41.00

"From the above composition it is seen that coca cola, that is, the syrup from which the drink is made, consists of about half sugar, one-third water, and the remainder the substances mentioned in the analysis above. In preparing the beverage about one ounce of the syrup, as indicated above, is used per glass. The active principle on which its vogue mainly depends is the caffein. When caffein is taken in from one- to two-grain doses, as is the case with those who drink tea, coffee and coca cola, the effects are prompt and, as a rule, agreeable. There is a sense of gentle exhilaration, a relief from the oppression of fatigue, and a decided wakefulness or stimulation of the nerve centers. This condition continues for several hours, so that the effect produced is never by any means transitory."

Many years have passed since this composition was submitted to the courts. Therefore, one should not conclude that this is the composition of present day Coca-Cola sirup.

Kola nuts are part of the seeds of a large tree. *Cola acuminata,* family *Sterculiaceae,* that is native to tropical western Africa. It has been and is cultivated in Brazil, the West Indies, and other tropical countries. The seeds, comprising two to five or more kernels, are contained in a pod which when mature contains one to ten seeds. They are about an inch in size and resemble the horse chestnut. They vary in size, however, with weights ranging from 1 ounce to about 0.2 ounce. When freshly picked, they are white or crimson, but change to brown or reddish-gray on drying and they turn a dark brown on roasting. Their odor is reminiscent of nutmeg. They are soft and fleshy when fresh, firm and woody after drying, and tough and solid after roasting.

Kola nuts contain caffeine and are used, for this reason, as the base of a stimulating beverage by natives of the Sudan. Some African natives use it as a food. Kola nuts are said to contain more caffeine than coffee and are used because of this for the manufacture of the kola beverages. Actually, however, published analyses of kola nuts give the caffeine content as 2.35, which, while considerably more than the caffeine content of coffee, that ranges from 1.2 to 1.8, is considerably less than that of tea which runs about 3.58 per cent.[5]

While extracts of kola nuts used in the preparation of kola flavors contain other components than caffeine, Beattie [6] stresses that these other components contribute little to the flavor. The main taste they give is one of a bitter character.

About the time that Wiley was writing his book on beverages (1919), the following formulation for a kola flavor was published. It is to be noted that this formulation contains no added caffeine and relies on the caffeine content of the fluid extract of kola used in the composition.

KOLA FLAVOR

Quinine hydrochloride	0.02 g
Citric acid	0.04 g
(dissolved in 10 g water)	
Orange-peel oil	4.00 drops
Sodium glycerophosphate	10.00 g
Fluid extract of kola	10.00 g
Saccharated iron oxide	15.00 g
Simple sirup	195.00 g

A completely different formulation is the one which follows. In this instance, the sugar is dissolved in the water to form a sirup in the customary manner and while it is kept boiling the caramel color is added and after cooling the other ingredients are added. The final flavored sirup is permitted to age at least 3 days.

KOLA SIRUP

Sugar	60.0 lb
Water	5.0 gal
Caramel color	40.0 oz
Alcohol	16.0 oz
Lime juice	16.0 oz
Vanilla extract	5.0 oz
Glycerol	4.0 oz
Kola flavor	4.0 oz

Another kola flavor found in the literature appears to require the addition of caffeine or a fluid extract containing caffeine. This is illustrative of the type that does not use kola nuts.

KOLA FLAVOR

Extract of coca leaves	2.0 dr
Neroli oil	6.0 dr
Lime oil	8 oz
Sweet orange oil	14.0 oz
Nutmeg oil	2.5 oz
Cassia oil	5.0 oz
Lemon oil	28.0 oz
Vanilla extract	28.0 oz
Alcohol (95%)	9.0 gal

Bennett [7] details a kola-type flavored sirup that contains many of the ingredients noted in the formulations detailed in this section. To make one gallon of extract 5.5 pounds of sugar are added to 4.5 pints of water to make a sirup in the customary manner. To the sirup are added 2 fluid ounces of caramel color, 0.5 ounce of citric acid dissolved in a little water, ¼ ounce of 85 per cent sirupy phosphoric acid, 1 ounce of the kola-type flavor, and 1 ounce of a modifying flavor which he terms Mellol.

KOLA-TYPE FLAVOR

1. Fluid extract of kola nuts	100 ml
Fluid extract of kola leaves	100 ml
Water	40 ml
Alcohol	40 ml
2. Lemon oil	100 ml
Alcohol	26 ml

Mix together 36 parts by volume of the first solution and 8 parts of the second solution and allow the mixture to stand 6 days. Use kieselghur as a filter aid and filter. One ounce is used to flavor the gallon of kola sirup.

KOLA-TYPE FLAVOR MODIFIER

Artificial oil of neroli	0.01 ml
Nutmeg oil	0.06 ml
Cassia oil	0.06 ml
Lime oil (distilled)	0.50 ml
Fluid extract of gambir	3.00 ml
Alcohol	1.00 oz

It can be seen from the preceding formulations that many kola flavors are modified. Beattie [6] discusses this aspect in considerable detail and gives several formulations of modified kola flavors. Representative of such compositions are the orange-kola, lemon-kola, and ginger-kola formulations that follow. In each of these soluble essences are used for the preparation of the kola-flavor modifier. It is customary to use 1 fluid ounce of an adequate kola-nut extract or tincture of kola nuts per gallon of sirup in conjunction with a similar volume of a blended kola flavor which, as can be seen, contains a blend of various flavors, but no kola. Other formulations employ from 1 fluid ounce to 6 fluid ounces per gallon.

ORANGE-KOLA FLAVOR MODIFIER

Essence	Parts
Sweet orange	25
Bitter orange	25
Mandarin	15
Curaçao	10
Vanilla	10
Bergamot	5
Petitgrain	2
Neroli	3
Sherry flavor (synthetic)	5

LEMON-KOLA FLAVOR MODIFIER

Essence	Parts
Green lemons	30
Sweet lemons	30
Limes	10
Lemongrass	5
Bergamot	5
Neroli	5
Sweet orange	5
Vanilla	5
Nutmeg	2
Cedrat	2
Rose	1

GINGER-KOLA FLAVOR MODIFIER

Essence	Parts
Ginger	60

Cassia	10
Vanilla	10
Gingergrass	5
Geranium	5
Bergamot	2
Capsicum	2
Cardamon	2
Grains of paradise	2
Pimento	2

The amount of caffeine in kola beverages (and thus in kola flavors) and whether the caffeine is a natural part of the kola extract employed or is added caffeine, obtained from tea, coffee, or other source, have been matters of considerable moment at various times. Thus certain States forbid the addition of caffeine to kola beverages (and flavors). Beattie [6] notes that Australian health experts comment adversely on finding as little as ¼ grain (16.2 mg) in a 7-ounce pack and that a United Kingdom editor suggested that ½ to 1 grain per pint equivalent to 32.4 to 64.8 mg per 8-ounce bottle be used. He also notes that another authority suggested a maximum of 0.65 grain (42 mg) per 8-ounce bottle. This is equivalent to the use of 8 ounces of caffeine per 50 gallons of kola sirup. W. P. Heath, a Vice President of the Coca-Cola Co., New York, wrote to the editor of the *Journal of the American Medical Association* (December 11, 1948 issue, volume 138, page 1132) and stated, ". . . . the caffeine content of Coca-Cola does not exceed 40 mg per bottle and is uniform everywhere in the world." In the United States the industry maximum is 0.5 grain (32.4 mg) per 6-ounce bottle.

c. **Ginger Ale**

Ginger ales are one of the most widely used carbonated-beverage flavors for not only are they consumed directly, but they are also widely used as mixers in prepared alcoholic drinks, like highballs.

The federal definition for ginger-ale flavor was given in Chapter 1. It is to be noted from this definition that ginger is an essential component of the flavor. Ginger ales can be placed into two main subgroups, which have to be taken into consideration in the manufacture of the flavor or of the flavored sirup, on the basis of sugar content, namely, ginger ales and as dry ginger ales. The second, as the word "dry" implies, contains less sugar than the first. At times a distinction is made on the basis of color also, namely, as golden ginger ale and as pale ginger ale but what interests us here is that the second type is usually also a "dry" type and is termed pale dry ginger ale.

(1) Ginger

There are many descriptions of ginger in the literature. One of the more concise is that of Worrell.[8] The dried rhizome of *Zingiber officinale*, Roscoe, a perennial plant is known as ginger. This plant resembles the common iris. Ginger has been used as a spice and in that sense a flavoring for centuries. It is probably indigenous to the countries of southern Asia. The variety grown in Jamaica is considered the best. Among the countries in which it is cultivated are Japan, China, India, Malaya, the Indo-Chinese States, the coastal regions of western Africa, and the East and West Indies.

Worrell says that the plant is grown from segments of the rhizome in moist, well-cultivated plots. The rhizomes are dug out when the overground portion begins to die, which usually occurs just about less than a year after planting. The procedure for preparing the rhizomes for market differs in various countries and places, but in general the treatment consists of partially or completely removing the outer cortical layers by peeling or scraping. The rhizomes are then washed and dried in the sun. At times they are bleached with chloride of lime or with sulfur dioxide before they are dried.

Ginger contains a volatile oil which gives it its characteristic aroma. This oil is present in about 3 per cent concentration. A number of pungent principles have been isolated from ginger. One of these has been named gingerol. According to Nelson,[9] gingerol is a very pungent yellow oil which has a specific gravity of 1.1073, a specific rotation of 12.9 degrees, a refractive index of 1.5212, and a methoxyl content of 9.26 per cent. Gingerol is readily decomposed into zingerone, a ketone, and enanthaldehyde. Zingerone has been identified as 1-(4-hydroxy-3-methoxyphenyl)-3-butanone. It has been synthesized from vanillin and acetone by reducing the resultant product with hydrogen and platinum to yield the ketone.

Zingerone is very pungent and it has a sweetish odor. It melts at 40° to 41°C. The homologues of zingerone, namely, the propyl, butyl, isobutyl, and *tert*-butyl derivatives, are all strongly pungent compounds. Related compounds, such as 2-hydroxystyryl phenyl ketone, $HOC_6H_4CH:CHCOCH_3$, 4-hydroxy-3-methoxystyryl phenyl ketone, $HOC_6H_3(OCH_3)CH:CHCOC_6H_5$, and similar ketones are also very pungent.

Ginger, in contrast to capsicum, loses its pungency when heated with alkaline hydroxides, while the pungency of capsicum is not destroyed by this treatment.

Shogaol is another pungent tasting substance which has been isolated from ginger. It is related to gingerol. It can be synthesized from zingerone and hexylaldehyde and has been shown to be 1-(4-hydroxy-3-methyoxyphenyl)-ethyl-1,2-heptenyl ketone.

(2) Ginger Standards

Standards for ginger were set by the Food and Drug Administration years ago. These were:

Ginger. The washed and dried, or decorticated and dried, rhizome of *Zingiber officinale,* Roscoe. It contains not less than 42 per cent of starch, not more than 8 per cent of crude fiber, and not more than 1 per cent of lime (CaO), not less than 12 per cent of cold-water extract, not more than 7 per cent of total ash, nor more than 2 per cent of ash insoluble in hydrochloric acid, nor less than 2 per cent of ash soluble in cold water.

Jamaica Ginger. Ginger grown in Jamaica. It contains not less than 15 per cent of cold-water extract, and conforms in other respects to the standards for ginger.

Limed Ginger, Bleached Ginger. Whole ginger coated with carbonate of calcium. It contains not more than 4 per cent of carbonate of calcium nor more than 10 per cent of total ash, and conforms in other respects to the standards for ginger.

The N.F. X recognizes only Jamaica, African, and Cochin gingers.

(3) Ginger Oil

The Essential Oil Association of America has promulgated the following tentative monograph for oil of ginger.

OIL OF GINGER

Oil of ginger is a volatile oil with the aromatic and persistent odor of ginger, but lacking the pungency or "bite" usually associated with ginger. It is widely used as a flavor in soft drinks, bakers' products and in candies. The dried rhizomes, from which the oil is made, derives its name from the country or locality of its origin, i.e., Jamaica, from the West Indies, Africa from the west coast of Africa, and both Cochin and Calicut

from southern India. The African and Cochin are the types most frequently used for distillation.
Botanical nomenclature—*Zingiber officinale,* Roscoe
Family—Zingiberaceae
Preparation—By direct steam distillation of the dried coarse-ground rhizome.
Physical and chemical constants—Color and appearance, light yellow to yellow; specific gravity at $15°/15°C$, 0.877 to 0.888; optical rotation $-28°$ to $-45°$; refractive index at $20°C$, 1.4880–1.4940.
Saponification Number—Not more than 20. Proceed as directed for the determination of esters (see Determinations Essential Oil Association of America No. 1A) using 5 grams of oil of ginger, accurately weighed. Calculate as follows
$$\frac{A \times 28.05}{B} = \text{Saponification No.}$$
 A is the result obtained by subtracting the number of milliliters of $0.5\ N$ hydrochloric acid required in the titration from the number of milliliters of $0.5\ N$ alcoholic potassium hydroxide originally taken. B is the weight in grams of the oil taken for analysis.
Descriptive characteristics—Stability: Alkali: Relatively stable to weak alkali. Unstable in the presence of strong alkali. Acids: Unstable in the presence of strong acids.
Solubility: Alcohol, soluble, usually with turbidity; benzyl benzoate, soluble in all proportions; diethyl phthalate, soluble in all protions; fixed oils, soluble in most fixed oils; glycerol, practically insoluble; mineral oil, usually soluble in all proportions; propylene glycol, practically insoluble.
Containers—Should be shipped in glass, tin-lined or aluminum containers.
Storage—Store preferably in tight, full containers in a cool place protected from light.

(4) Flavors

Ginger-ale flavors are like root-beer flavors in that they are made from extracts of ginger and of plants or from essential oils. Years ago fluid extract of ginger was included as an item in the United States Pharmacopeia. A preparation about four times the volume of the U.S.P.X formulation can be made by moistening 8 pounds of Jamaica ginger powder with a portion of a gallon of 85 per cent alcohol and packing this mixture lightly in a narrow percolator. When the liquid begins to drip, the drain port is closed or stoppered, the remainder of the 85 per cent alcohol is used to cover the ginger powder, and the mass is allowed to steep for several days, after which percolation is permitted to take place in the customary manner. N.F.X (1955) gives a somewhat similar description of the preparation, in effect, obtaining the extractive matter from 1000 grams of ginger in 1000 milliliters of fluid extract.

A ginger extract can be made from ginger oleoresin. This contains not less than 18 milliliters and not more than 35 milliliters of volatile oil of ginger in each 100 grams of the oleoresin. This is made by extracting coarse ginger powder with acetone, ethyl alcohol, or ether by percolation, recovering the major portion of the volatile solvent by distillation and allowing the remainder of the volatile solvent to escape by evaporation.

A soluble ginger extract [23] which can be used as the base of both golden and pale dry types of ginger ales comprises:

SOLUBLE GINGER EXTRACT

Oleoresin ginger	3 avoir. oz
Ginger oil	100 minims
Alcohol	64 fl oz
Water, sufficient for	1 gal

This soluble ginger extract is prepared by dissolving the oleoresin ginger and the oil of ginger in about 60 fluid ounces of the alcohol, reserving 4 ounces of the alcohol to assist in clearing the extract at the final step. The water, which will amount to about 64 fluid ounces, is added gradually with thorough agitation during a period of several hours. One ounce of a filter aid, such as purified diatomaceous or Fuller's earth, is added per gallon of extract if more than 1 gallon is being prepared and is mixed thoroughly. This stirring is continued for 1 hour and then the mixture is permitted to stand until the filter aid settles out. The mixture is stirred again to resuspend the filter aid and it is again allowed to stand to settle. This alternate stirring and settling is continued for a day to make certain that all insoluble matter is trapped. The mixture is then allowed to stand overnight and is filtered through paper or with the aid of a mechanical filter, if a large batch is being prepared. It is good practice to coat the filter initially with some of the filter aid wetted with some of the extract. The initial filtrate is returned to the main preparation and refiltered. Filtration is continued until the extract is sparkling clear.

The reserved ratio of 4 fluid ounces of alcohol is added to the final filtrate to dilute the saturated ginger solution somewhat so that none will precipitate with variation in temperature or because of a slight amount of evaporation of the alcohol. The final volume of the extract can be adjusted to 1 gallon or other proportionate volume to make up for loss by evaporation during filtration. For instance, if the final filtrate is shy 6 fluid ounces to make 1 gallon, then 3 fluid ounces of alcohol is added and stirred into the extract and then 3 fluid ounces of water is added and stirred into the extract. It is necessary to make certain that the mixture is thoroughly mixed and that no ginger constituents are reprecipitated.

It will be clear from an examination of the following ginger-ale flavors that nearly all comprise a mixture of a ginger extract and citrus oils. In addition various components, such as rose, cinnamon, vanilla, etc., are added to give the flavor a bouquet or top note and cinnamon, clove, cardamom, etc., to give it added spice character, and capsicum, pimento, etc., to give added bite.

Examples of ginger-ale flavor extracts made from oleoresin ginger are the following:

GINGER-ALE FLAVOR EXTRACT

Ginger oleoresin	24.0 oz
Capsicum oleoresin	5.5 oz
Terpeneless lemon oil	36.0 oz
Terpeneless orange oil	12.0 oz
Cinnamon oil	1.5 dr
Rose oil	0.5 dr
Clove oil	1.5 dr
Alcohol	5.5 gal
Water	3.0 gal

Two ounces of this flavor extract are used for each gallon of simple sirup.

GINGER-ALE FLAVOR EXTRACT

Ginger oleoresin	15.0 oz
Capsicum oleoresin	2.0 oz
Lemon extract	5.0 pt
Orange extract	2.5 pt
Alcohol	1.3 gal
Water	0.7 gal

Three ounces of this flavor extract are used for each gallon of simple sirup.

Bennett [7] details a formulation for the preparation of a pale dry ginger-ale extract. Four ounces of ginger oleoresin, 2 ounces of lemon oil, and 2 ounces of distilled lime oil are dissolved in 0.5 gallon of 95 per cent alcohol and to this mixture 0.5 gallon of distilled water is added slowly, with constant stirring. The mixture is permitted to stand for a few days after which the extract is drawn off leaving any precipitate behind. A filter aid is added to the liquid and stirred in and the mixture is filtered. Four additional ounces of 95 per cent alcohol are added to keep the oils in solution, i.e., to prevent precipitation if the temperature is lowered, and to assist in dissolving the extract in the sirup. About 3.5 ounces of the extract are used for each gallon of simple sirup.

Illustrative of ginger-ale flavors in which only the oils are used are the following:

GINGER-ALE FLAVOR

Ginger oil	4.0 oz
Capsicum oil	1.0 oz
Lemon extract	16.0 oz
Orange extract	8.0 oz
Alcohol	3.5 pt
Water	3.5 pt

GINGER FLAVOR

Ginger oil	12 oz
Amyl butyrate	30 oz
Ethyl enanthate	20 oz
Ethyl acetate	20 oz
Pettigrain oil	4 oz
Cardamom oil	4 oz
Clove oil	2 oz
Galanga oil	12 drops
Alcohol	16 oz

Representative of a ginger-ale flavored sirup is the following composition:

GINGER-ALE FLAVORED SIRUP

Soluble ginger extract	8.0 oz
Soluble orange extract	16.0 oz
Soluble lemon extract	8.0 oz

Vanilla extract	2.0 oz
Caramel color	2.5 oz
Citric acid solution	32.0 oz
Sugar	46.0 lb
Water	6.5 gal

The sirup is made in the manner detailed in Chapter 2 and the flavored sirup can be prepared as described in the preceding chapter. One to $1\frac{1}{4}$ ounces of the final ginger-ale flavored sirup is used per 8-ounce soda bottle.

The soluble ginger extract, whose preparation was detailed on page 139, can be used to make various ginger-ale flavors which are considered appropriate for the flavoring of non-nutritive sweetened carbonated beverages.[23]

GINGER-ALE FLAVORS

Ingredient	Pale dry fl oz	Pale dry fl oz	Pale dry fl oz	Golden fl oz	Golden fl oz
Soluble ginger extract	128	128	128	128	128
Terpeneless lemon oil	$\frac{3}{8}$	$\frac{1}{8}$	$\frac{1}{4}$	$\frac{1}{4}$	$\frac{1}{8}$
Terpeneless lime oil	$\frac{1}{8}$	$\frac{3}{8}$	$\frac{1}{4}$	$\frac{1}{4}$	$\frac{3}{8}$
Alcohol	40	40	40	40	40
Water, sufficient for	192	192	192	192	192

The lemon and lime oils are dissolved in the alcohol and then sufficient water is added to make the volume $\frac{1}{2}$ gallon. The soluble ginger extract is added gradually to this alcoholic solution of citrus oils. Ordinarily the final solution mixture will be clear, but if it is not, it can be filtered through paper precoated with a filter aid.

To prepare the pale dry ginger-ale bottling concentrate, 1.5 ounces of the pale dry formulations, 3 fluid ounces of 50 per cent citric acid, sufficient caramel coloring, and 1.5 ounces of a calcium cyclamate, such as Sucaryl calcium, are added to enough water to make a final volume of 1 gallon. In manufacturing the beverage, 1 ounce of the bottling concentrate is used in a 7-ounce bottle and carbonated water containing 4 volumes of carbon dioxide is added. This beverage will then contain 0.16 per cent of calcium cyclamate.

To prepare the golden ginger-ale bottling concentrate, 1.5 fluid ounces of the golden formulations, 2 fluid ounces of 50 per cent citric acid, sufficient caramel coloring, and 1.75 ounces of a cyclamate, such as Sucaryl calcium, are added to enough water to make a final volume of 1 gallon. In manufacturing the beverage, a throw of 1 ounce is used per 7-ounce bottle with carbonated water containing 4 volumes of carbon dioxide. This carbonated beverage will contain 0.19 per cent of calcium cyclamate.

d. **"Cream" Flavor**

Together with the other flavors of this group of specialty flavors, cream-soda-water flavor was defined in Chapter 1. The principal flavoring ingredients are vanilla or vanillin. Formerly coumarin was a major flavoring ingredient, but since the discovery that coumarin may be toxic to human beings [10] and its removal as an optional ingredient in the Food and Drug Administration definitions and standards of certain foods, it cannot be considered acceptable in carbonated soft drinks.

It is beyond the scope of this book to discuss in great detail the preparation of vanilla extract. There are two principal methods used for its preparation. The first and the oldest is maceration. The second method and that of greater commercial importance is percolation.

The maceration method is relatively simple. It lends itself to the preparation of vanilla extract even at the household level. In simplest terms, chopped vanilla beans are extracted by soaking in an aqueous alcohol mixture for a relatively long time.

A specific example follows. Cut an ounce of vanilla beans finely, preferably with the aid of a machine. Allow the cut beans to soak in 10 ounces of a mixture of equal volumes of 95 per cent alcohol and water for a few days. After the beans appear to be exhausted, decant the supernatant liquid and transfer the extract to containers for use as needed.

The percolation or displacement method is used more widely than the maceration procedure. It lends itself to the more rapid preparation of a flavoring extract and to the possible establishment of a continuous process for the preparation of vanilla extract as contrasted to the usual batch process. In general, this will not be performed by the soft drink manufacturer. If the extracting solvent is kept warm, that is, the temperature is kept at 110° to 115°F, the extraction will be more complete in a relatively shorter period of time, but higher temperatures are not desirable, because there will be both loss of flavor and loss of alcohol through volatilization.

The following is an example of a relatively rapid method [11] for the manufacture of vanilla extract. Place the vanilla beans, after being weighed, in a chopper and disintegrate them thoroughly. Transfer the chopped beans to a perforated-metal vessel, resembling in some measure the one used in a coffee percolator, and place it in a percolator. Pour the washes tabulated as follows through the chopped beans, and then mix, filter, and store the mixture in a tank, or in containers. The vanilla extract contains 30 per cent alcohol when 95 per cent alcohol is used.

VANILLA EXTRACT

First Wash

Vanilla beans	137 lb
Alcohol	50 gal
Water	20 gal

Second Wash

Alcohol	6 gal
Water	20 gal

Third Wash

Water	37 gal

Fourth Wash

Simple sirup	28 gal

The total ingredients are thus 137 pounds of vanilla beans, 56 gallons of 95 per cent alcohol, 28 gallons of simple sirup, and 77 gallons of water. A somewhat weaker vanilla extract can be prepared by diluting the 150 gallons made in the manner detailed with 50 gallons of 30 per cent alcohol.

Jacobs [12] discussed in considerable detail the preparation of imitation vanilla extracts. Virtually all of these contained coumarin as a component, consequently all of these must be modified by the omission of coumarin and the substitution of other ingredients. It must be stressed that coumarin is not a normal constituent of vanilla beans and consequently is not a component of true vanilla extract.

In general, imitation vanilla extracts consist of a solution of vanillin, a modifying flavoring component, such as a coumarin substitute, piperonal (heliotropin), bourbonal (so-called ethyl vanillin, a name which is incorrect since it is in reality a methyl derivative of vanillin), and solvents, such as glycerol, alcohol, and water. White imitation vanilla extracts contain no caramel, whereas ordinary imitation vanilla extracts usually are colored with caramel.

To prepare an imitation vanilla extract, the vanillin and the other flavoring components are dissolved in the alcohol. The sugar or sirup, the glycerol if used, and the caramel solution are dissolved in the water and both solutions are mixed together. The final mixture is allowed to stand a few days before filtering and bottling or use in making the flavored sirup.

A typical "cream" soda flavor extract is the following:

"CREAM" SODA FLAVOR EXTRACT

Vanillin	20 oz
Coumarin substitute	8 oz
Alcohol	2 gal
Glycerol	1 gal
Water	1 gal

One ounce of this preparation will flavor 5 gallons of sirup.

It is important to remember when making concentrated vanillin solutions that vanillin and the compounds used to modify the flavor of vanillin are not very soluble. Consequently, when lesser amounts of alcohol are used, there is danger that some of the compounds will precipitate from solution when the temperature drops.

VANILLA FLAVOR RATIOS

1 part vanilla beans	≈ 0.07 part vanillin
1 pound vanilla beans	≈ 1.125 oz vanillin
standard vanilla extract	≈ 0.7% vanillin solution
1 part piperonal (heliotropin)	≈ 2 parts vanillin
1 part bourbonal	≈ 3 to 4 parts vanillin
(so-called ethyl vanillin)	
1 part propenyl guaethol	≈ 16 to 25 parts vanillin
1 part 6-methylcoumarin	≈ 4.5 to 5 parts vanillin

e. **Quinine Water**

A specialty-flavored beverage that has had considerable vogue in Great Britain and has had some popularity in the United States recently is quinine water. In the Soft Drinks Minimum Standard

(Food Standards [Soft Drinks] Order, 1953, 1828) of Great Britain which came into effect on December 20, 1953, the standards that had been in force for "Indian and Quinine Tonic" were continued. These standards required that there is a minimum of 1 pound 2 ounces of sugar per 10 gallons, a maximum of 82 grains of saccharin per 10 gallons, and a minimum of 0.5 grain of quinine (calculated as quinine sulfate) per pint. These standards should prove of assistance in the formulation of a flavored sirup for the manufacture of this type of specialty-flavored beverage.

Another quinine water or tonic formulation contains 8 grains of quinine sulfate in a mixture of 4 pints of carbonated lemon soda and 4 pints of carbonated water, that is, 1 grain of quinine sulfate per pint of finished beverage.

2. FRUIT FLAVORS

Fruit flavored carbonated beverages comprise a substantial portion of the soft-drink market as noted in Chapter 1. In the preceding chapter, the use of true fruit flavors was discussed. In this chapter, artificial fruit flavors will be considered and in Chapter 8, flavor emulsions of both the true fruit, principally citrus essential oil type, and artificial flavor type will be discussed in detail.

It would not be appropriate in a book of this scope to discuss the formulation of artificial fruit flavors in very great detail. This has been done by Jacobs.[13] Artificial fruit-flavor formulations vary from simple compositions containing a few flavoring components to very complex formulations. In each instance in the following discussion, both simple and more complex formulations will be detailed, but these are only illustrative. It is difficult to give specific directions for the use of artificial flavors for carbonated beverages. Such use depends in great measure on the flavor strength of the preparation and depending on this flavor strength, from as little as 1/20 of an ounce per gallon of sirup to the customary 1 ounce to the gallon is employed. The more common ratios are $1/8$ to $1/2$ ounce to the gallon of a beverage sirup.

a. Strawberry Flavor

Studies of natural strawberry flavor [14,15,16] disclose that it contains many components. Artificial strawberry flavors have been prepared with ingredients which are present in the natural product as well as ingredients which are wholly synthetic. In the second group are the strawberry "aldehydes," so-called Aldehyde C_{16}, namely, ethyl phenylglycidate and ethyl methylphenylglycidate. Examples of both types are given. Formulas suggested by Walter [17] follow.

STRAWBERRY FLAVOR

(Cultivated Berries)

Amyl butyrate	1.5 lb
Benzyl acetate	1.5 lb
Ethyl acetate	1.0 lb
Ethyl butyrate	1.0 lb
Methyl salicylate	0.5 lb
Amyl butyrate	0.5 lb
Ethyl benzoate	6.0 oz
Nerolin	4.5 oz
Amyl formate	4.0 oz
Ceylon cinnamon oil	12.0 dr
Coumarin substitute	4.0 dr

STRAWBERRY FLAVOR

(Wild Berries)

Amyl acetate	3.0 lb
Benzyl acetate	1.5 lb
Ethyl acetate	1.5 lb
Ethyl butyrate	0.5 lb
Methyl salicylate	0.5 lb
Ethyl benzoate	6.0 oz
Nerolin	5.0 oz
Coumarin substitute	1.0 oz
Vanillin	12.0 dr
Ceylon cinnamon oil	6.0 dr

Some 15 years ago, Blumenthal [18] gave the following formulation:

STRAWBERRY FLAVOR

	ounces
Yara yara (10% solution in benzyl benzoate)	1.00
Amyl acetate	3.00

Aldehyde C_{14}	1.50
Iris concrete (10% solution)	0.75
Bitter-almond flavor essence	1/20
Rose essence	1/20
Jasmine essence	1/100
Ionone	1/100
Bourbonal	1/12
Alcohol	16

An example of a formulation in which all of the components are synthetic compounds is given by Jacobs.[13]

STRAWBERRY FLAVOR

Isoamyl acetate	440
gamma-Undecalactone	220
Benzyl benzoate	134
Ethyl acetate	100
Bourbonal	40
Benzaldehyde	40
Ionone	14
Geraniol	12

This formulation may then be diluted 1 volume to 3 of ethyl alcohol.

The influence that more modern thought has had on the formulation of strawberry flavors can be seen in the following strawberry formulation suggested by Benezet.[19] In this formulation an attempt is made to incorporate into the composition those compounds that were isolated from natural strawberries by Coppens and Hoejenbos, and the components of the older formulations given in the preceding paragraphs. Modifications of the Benezet formulation should include information gleaned from the work of McGlumphy [15] and of Dimick and his coworkers.

STRAWBERRY FLAVOR

Ethyl methylphenylglycidate	400
Benzylidene isopropylidene acetone	100
Ethyl acetoacetate	80
Ethyl butyrate	50
p-Cresyl butyrate	50
Ethyl acetate	30
Ethyl caproate	20

Phenethyl butyrate	20
Yara yara	20
Vanillin	20
Ethyl phenylacetate	20
beta-Naphthol butyrate	20
Essence of Portugal	20
Methyl naphthyl ketone	20
Maltol	14
Ethyl benzoate	10
Ethyl cinnamate	10
Nerol	10
Amyl acetate	10
Hexyl acetate	10
Coumarin substitute	10
Essence of iris concrete	10
Essence of neroli	10
Borneol	5
Borneol acetate	5
Methyl anthranilate	5
Ethyl salicylate	5
Hexyl alcohol	5
Terpineol	5
Biacetyl	2
Pentanal	2
Isojasmone	2

b. **Raspberry Flavor**

Natural raspberry flavor is also a very complex mixture as shown by a number of investigators.[20] As with strawberry flavor formulations, raspberry flavor formulas vary from relatively simple compositions containing six or seven ingredients to very complex ones having over thirty ingredients. Such formulations vary from those which rely almost completely on natural oils to those which depend almost completely on synthetic chemicals. An illustration of the first is given by Walter.[17]

RASPBERRY FLAVOR

Isobutyl acetate	3.0 lb
Amyl acetate	2.0 lb
Ethyl acetate	1.5 lb
Ethyl formate	0.5 lb
Benzyl benzoate	6.0 oz

Orris oil (10% solution)	5.0 oz
Vanillin	3.0 oz
Petitgrain oil	12.0 dr
Clove oil	12.0 dr
Ceylon cinnamon oil	6.75 dr
Jasmine oil (10%)	1.125 dr
Rose oil	0.125 dr

Blumenthal suggested a formulation in which reliance on natural oils is still marked.

RASPBERRY FLAVOR

Phenethyl alcohol	50.0 oz
Benzyl acetate	40.0 oz
Peru balsam	10.0 oz
Propyl acetate	10.0 oz
Clove oil, terpeneless	5.0 oz
Cinnamon oil	5.0 oz
Amyl acetate	2.0 oz
Bitter almond oil	2.0 oz
Rose oil	2.0 oz
Neroli, synthetic	2.0 oz
Orris (1% solution)	1.0 oz
Vanillin	1.0 oz
Ethyl acetate	0.5 oz

This mixture was dissolved in 500 ounces of ethyl alcohol to make the essence. Correspondingly greater volumes of essence must be used to flavor the sirup than if the ester mixture alone is used.

An example of a formulation in which all of the components are synthetic compounds is given by Jacobs.[13]

RASPBERRY FLAVOR

Isobutyl acetate	380
Isoamyl acetate	250
Ethyl acetate	200
Ethyl formate	60
Benzyl benzoate	50
Vanillin	20
Eugenol	15
Linalool	15
Cinnamaldehyde	5
Ionone	3
Geraniol	1
Benzyl acetate	1

SPECIALTY AND FRUIT FLAVORS 151

The essence can be made from this formulation by diluting with 9 volumes of ethyl alcohol. Comparison with the preceding formulations shows that some of the oils have been replaced by their isolates as, for instance, clove oil by eugenol and cinnamon oil by cinnamaldehyde. Another such formulation is given on page 124. The influence of more modern thought on raspberry-flavor formulation can be seen by comparing the following formulation suggested by Benezet [19] with the others given in this section.

RASPBERRY FLAVOR

Ethyl methylphenylglycidate	400
Benzylidene isopropylidene acetone	100
Methoxyacetoxyacetophenone	60
Benzyl acetate	50
Phenethyl alcohol	50
Essence of Portugal	50
Isobutyl acetate	40
Vanillin	30
Methylionone	25
beta-Ionone	25
Coumarin substitute	10
Iris concrete essence	15
Ethyl acetate	10
Ethyl caproate	10
Isoamyl caproate	10
Hexanyl acetate	10
Hexenyl acetate	10
Methyl salicylate	10
Ethyl benzoate	10
Methyl butanol	10
Bornyl salicylate	10
Essence of clove	10
Essence of geranium	10
Hexyl alcohol	5
Hexenol	5
Anisaldehyde	5
Benzaldehyde	5
Acetylmethylcarbinol	3
Biacetyl	2

In the Benezet formulation, an attempt is made to incorporate into the composition those compounds which were isolated by Coppens and Hoejenbos and also Bohnsack.[20] Benezet suggests that in addition to the compounds mentioned, one might also use some of the cyclohexanol esters, some rhodinol, and some borneol. He also recommends the use of a natural concentrate of raspberry juice.

c. **Cherry Flavor**

In a manner analogous to those given for strawberry and for raspberry, one can give old and new formulations for cherry flavor. Thus Walter suggested the following:

CHERRY FLAVOR

Ethyl acetate	3.5 lb
Amyl formate	2.25 lb
Ethyl enanthate	12.0 oz
Amyl alcohol	8.0 oz
Benzaldehyde	7.0 oz
Vanillin	2.5 oz
Benzyl benzoate	2.5 oz
Clove oil	2.0 oz
Cognac oil	1.5 oz
Cinnamon oil	0.5 oz

A similar type of formula for a cherry-flavor essence was proposed by Blumenthal.

CHERRY FLAVOR

Ethyl acetate	64.0 oz
Ethyl enanthate	20.0 oz
Ethyl alcohol	16.0 oz
Benzyl benzoate	8.0 oz
Clove oil	6.0 oz
Benzaldehyde or bitter almond oil	4.5 oz
Ethyl pelargonate	2.0 oz
Cinnamon oil	2.0 oz
Cognac oil	1.5 oz
Amyl alcohol	1.5 oz
Vanillin	1.0 oz
Neroli oil, synthetic	0.125 oz

Again we can refer to Benezet for a formulation which takes into consideration some of the modern developments. Thus a characteristic of cherry flavor is the presence of benzaldehyde. Related aldehydes like *p*-tolualdehyde can also carry this effect and compounds like ethyl acetoacetate and acetoin, that is, acetylmethylcarbinol, help in producing a natural effect.

CHERRY FLAVOR

Ethyl acetate	408
Ethyl acetoacetate	100
Ethyl enanthate	100
Ethyl caprate	100
Isoamyl butyrate	50
Ethyl caproate	50
Ethyl caprylate	50
Benzaldehyde	50
Bitter almond essence, F.F.A.	30
p-Tolualdehyde	20
Vanillin	10
Terpenyl butyrate	5
Geraniol	5
Geranyl butyrate	5
Essence of cloves, Bourbon	5
Essence of cinnamon	5
Acetylmethylcarbinol	5
Essence of wine dregs	2

d. **Grape Flavor**

There are two principal types of grape flavor, (1) the "winey" or California or European type of grape and (2) the Eastern American or Concord type of grape flavor. In the first, ethyl enanthate (a compound which is called by a number of other names, such as enanthic ether, oenanthic ether, ethyl heptoate, ethyl heptanoate, ethyl heptylate) is a principal ingredient and in the second, methyl anthranilate and other anthranilates are common components. The Concord type of flavor is the one more commonly used for carbonated beverages. Stetson [21] details one of such formulations.

GRAPE FLAVOR

Port wine	75.0
Grape juice	50.0
Benzyl butyrate	10.5
Methyl anthranilate	4.5
Methyl salicylate	0.5

Fluid extract of valerian 3.0
Glycerol 25.0
Alcohol (95%) 150.0

An attempt to incorporate substances isolated from grape juice is shown in the formulation from Benezet.[22]

GRAPE FLAVOR

Ethyl enanthate	200
Ethyl acetate	150
Methyl anthranilate	100
Ethyl palmitate	100
Ethyl acetoacetate	100
Ethyl laurate	88
Ethyl formate	50
Ethyl caprylate	50
Ethyl pelargonate	50
Ethyl butyrate	20
Ethyl caproate	20
Ethyl valerate	20
Ethyl cinnamate	20
Geranyl butyrate	10
Geraniol	5
Essence of dregs of wine	5
Acetylmethylcarbinol	3
Essence of muscat nut	3
Biacetyl	2
Essence of sage	2
Essence of cardamom	2

LITERATURE CITED

1. M. B. Jacobs, *Am. Perfumer Essent. Oil Rev.*, **50**, 565 (1947); **51**, 55 (1948).
2. *New Era Formulary*, New York, D. O. Haynes & Co., 1893.
3. *The National Formulary*, 10th Ed., N.F. X, Am. Pharm. Assoc., Philadelphia, Lippincott, 1955.
4. H. W. Wiley, *Beverages and their Adulteration*, Philadelphia, Blakiston, 1919.
5. W. H. Ukers, in M. B. Jacobs, Ed., *Chemistry and Technology of Food and Food Products*, Vol. II, 2nd Ed., New York, Interscience Publishers, 1951.
6. G. B. Beattie, *Perfumery Essent. Oil Rec.*, **47**, No. 12, 437 (1956).

7. H. Bennett, Ed., *The Chemical Formulary*, Vol. IV, New York, Chemical Publishing, 1939.
8. L. Worrell, in M. B. Jacobs, Ed., *Chemistry and Technology of Food and Food Products*, Vol. II, 2nd Ed., New York, Interscience Publishers, 1951.
9. E. K. Nelson, *J. Am. Chem. Soc.*, **39**, 1466 (1917).
10. M. B. Jacobs, *Am. Perfumer Essent. Oil Rev.*, **62**, 53 (1953).
11. M. B. Jacobs, *Am. Perfumer Essent. Oil Rev.*, **48**, No. 2, 59 (1946).
12. M. B. Jacobs, *Am. Perfumer Essent. Oil Rev.*, **48**, No. 3, 56 (1946).
13. M. B. Jacobs, *Synthetic Food Adjuncts*, New York, Van Nostrand, 1947.
14. A. Coppens and L. Hoejenbos, *Rec. trav. chim. Pays Bas*, **58**, 680 (1939).
15. J. H. McGlumphy, *Food Technol.*, **5**, 353 (1951).
16. K. P. Dimick and B. Makower, *Food Technol.*, **10**, 73 (1956).
17. E. Walter, *Manual for the Essence Industry*, New York, Wiley, 1916.
18. S. Blumenthal, *Food Manufacturing*, New York, Chemical Publishing, 1942.
19. L. Benezet, *La Parfumerie Moderne*, **43**, No. 22, 61 (1951); *Am. Perfumer Essent. Oil Rev.*, **60**, 267 (1952).
20. M. B. Jacobs, *Am. Perfumer Essent. Oil Rev.*, **67**, No. 1, 49 (1956).
21. R. Stetson, *Am. Perfumer Essent. Oil Rev.*, **67**, No. 1, 58 (1956).
22. L. Benezet, *Am. Perfumer Essent. Oil Rev.*, **62**, 429 (1953).
23. "*Sucaryl*" *Sweetened Beverages*, Technical Information No. 42, North Chicago, Abbott Laboratories, 1957.

SELECTED BIBLIOGRAPHY

The Givaudan Flavorist, No. 5 (1957).

M. B. Jacobs, *Synthetic Food Adjuncts*, New York, Van Nostrand, 1947.

S. Blumenthal, *Food Manufacturing*, New York, Chemical Publishing, 1942.

E. Walter, *Manual for the Essence Industry*, New York, Wiley, 1916.

C. A. Nowak, *Non-Intoxicants*, St. Louis, C. A. Nowak Publisher, 1922.

"*Sucaryl*" *Sweetened Beverages*, Technical Information No. 42, North Chicago, Abbott Laboratories, 1957.

Bottlers Yearbook, Surrey, England, BYB Limited, 1956.

Chapter 8

EMULSIONS AND SPECIALTIES

Many carbonated beverages are made with flavor emulsions. Such flavors were originally prepared as substitutes for alcoholic flavoring extracts, but they need no longer be considered as substitutes. Jacobs [1] pointed out that emulsion flavors have several advantages, namely, (1) they are much less expensive to prepare than alcoholic extracts since a principal component is water; (2) there is no loss of volatile solvent to be considered in such formulations; (3) they are generally prepared in more concentrated form, consequently they usually occupy less storage space than alcoholic extracts and the filling, packing, and labor costs are lower because their bulk is less; (4) duplication, standardization, and quality control of the product are simplified; and (5) the loss by volatilization of many of the flavor components themselves is diminished. In addition to these advantages, the manufacture of stable cloudy, ringless beverages is made easier by use of certain types of flavor emulsions. These will be considered in a subsequent section of this chapter.

There are a number of disadvantages of emulsions. (1) Since they contain no alcohol, they do not have the extra lift that alcohol gives to the aroma; (2) they must be properly prepared, otherwise they settle out or separate; (3) it is often imperative to include a preservative to prevent spoilage, decomposition, or separation.

A number of emulsifying agents are available for making food emulsions. Among these are the water-soluble gums, principally, gum arabic, gum tragacanth, and gum karaya. In more recent years guar, algin and sodium alginate, methyl cellulose, sodium celluloseglycolate, and demethoxylated pectins have been tried. These and many others have been considered in detail by Jacobs.[1] In the preparation of beverage flavor emulsions, gum arabic, also known as gum acacia, is considered the emulsifying agent of choice. Gum arabic is the dried gummy exudation obtained from the stems and branches of trees of the genus *Acacia*, particularly species like *A. senegal*, *A. arabica*, and *A. verek*. Most of the gum arabic is supplied by the Sudan. There are two principal types of gum arabic that are known as hashab and talha, the first constituting about 90 per cent of the gum coming from Sudan. Gum arabic is obtained by cutting a strip ranging from 8 to 12 inches in length by 1/8 inch deep into the main branches of a tree during the months of October or January. The gum oozes out, hardens on the tree, and is removed after about 40 days. This is repeated, the yield diminishing after the first collection which gives about 35 to 40 per cent of the total gum.

Since in many formulations, the gum is used directly without prior solution and filtration, the best quality gum free from extraneous matter should be purchased.

1. FRUIT-FLAVOR EMULSIONS

A simple and general method for the preparation of fruit-flavor emulsions is the following. To make 1 gallon of flavor emulsion, place 5 ounces of gum-arabic powder in a mixing vessel, wet it with 2 fluid ounces of an artificial fruit-flavor oil or a fortified fruit-flavor oil of equivalent flavor power and mix at high speed with 8 ounces of 32°Bé sirup. Continue stirring for about 10 minutes until a smooth mass is obtained. Add while the stirring is continued 108 fluid ounces of water to which have been added 2 ounces of certified color, 5 fluid ounces of caramel coloring, and 5 grams of sodium benzoate, to act as a preservative. The entire mixture is stirred for about 20 minutes and is then passed through a homogenizer twice.

Such preparations are 2 ounce to the gallon flavors, that is, 2 ounces of the flavor emulsion are used for the flavoring of each gallon of beverage sirup. They produce clear sparkling beverages. The strength of flavor may be varied by using more or less of the flavor oil.

For strawberry, cherry, and raspberry flavors, use certified amaranth (FD&C Red No. 2). For grape, use a certified red color mixture, such as a commercial shade B.

In making these emulsions it is important to have the oil wet the gum arabic thoroughly. It is also important to remember not to run the emulsion through the homogenizer too often, for this treatment may break the emulsion instead of stabilizing it.

An example of an imitation grape flavor of the emulsion type [2] which is suggested for use in beverages sweetened with non-nutritive sweeteners is the following:

IMITATION GRAPE-FLAVOR EMULSION

Methyl anthranilate	1/3 fl oz
Cognac oil	1/8 fl oz
Ethyl acetate	1/8 fl oz
Ethyl butyrate	1/8 fl oz
Glacial acetic acid	20 minims
Balsam of Peru oil	20 minims
Gum arabic	4 oz
FD&C Red No. 2	2.5 oz
FD&C Blue No. 1	0.25 oz
Propylene glycol or glycerol	1 pt
Grape juice, sufficient for	1 gal

The methyl anthranilate, cognac oil, ethyl acetate, ethyl butyrate, glacial acetic acid, and balsam of Peru oil are mixed together and the mixture is added to the gum arabic which is rubbed thoroughly so that the gum is wetted by the oils. One pint of grape juice is added, the mixture is stirred again, and then is run through a homogenizer or colloid mill so that the particle size is reduced and a good dispersion is obtained. The FD&C colors are dissolved in about 0.5 gallon of the grape juice and to this color solution the propylene glycol or glycerol is added. The emulsion initially prepared is added to the color solution while stirring and after sufficient grape juice is added to make a total volume of 1 gallon; the final mixture is run through a homogenizer or colloid mill to emulsify the entire batch.

To prepare a bottling concentrate, 2 ounces of the imitation grape-flavor emulsion, 2 fluid ounces of 50 per cent citric acid, and 2.5 ounces of a cyclamate, like Sucaryl calcium, are added to enough

EMULSIONS AND SPECIALTIES

water to make 1 gallon. In manufacturing the beverage, a throw of 1 ounce of the bottling concentrate is used per 7-ounce bottle, with the carbonated water containing 1.5 to 2.0 volumes of carbon dioxide. The carbonated beverage produced will contain 0.27 per cent of calcium cyclamate.

2. CLEAR BEVERAGE CITRUS-OIL EMULSIONS

Some soft drink consumers prefer clear rather than cloudy carbonated soft drinks. Emulsions can be prepared to yield this type of product. The method of preparation is similar to that described for the preparation of fruit-flavor emulsions.

A typical clear lemon-lime flavor emulsion consists of ingredients in the following proportions: 0.5 ounce of terpeneless lemon oil, 10 drops of terpeneless lime oil, 5 ounces of gum arabic, and 8 ounces of 32°Bé sirup that are mixed together and then are mixed with sufficient water containing about 0.5 ounce of tartrazine (FD&C Yellow No. 5) to make one gallon of total mixture. The final mixture is passed through a homogenizer. Terpeneless oils, which are explained in Chapter 5, section 4.a, must be used in such preparations in order to make certain that the final drink will be clear. If the terpene-bearing oil is used, the low solubility of the terpenes will cause them to be precipitated on dilution and thus yield a beverage of poor appearance.

3. SPECIALTY-FLAVOR EMULSIONS

Root-beer flavor, sarsaparilla flavor, and birch-beer flavor emulsions can be made in a manner similar to that detailed for fruit-flavor emulsions. Typical formulations contain 4 to 6 ounces of the flavor oil, 6 ounces of gum arabic, and 8 ounces of 32°Bé sirup which are mixed together as mentioned. Sufficient caramel coloring, usually about 2 ounces, and 5 grams of sodium benzoate are dissolved in 112 fluid ounces of water and this is mixed into the gum-oil-sirup mass. The entire mixture is then passed through a homogenizer. Two ounces of the flavor are used per gallon of sirup.

Imitation vanilla flavor can be made in a similar manner. Heat 2 pounds of propylene glycol to 110°F and dissolve 2.25 ounces of bourbonal (so-called ethyl vanillin) and 0.5 ounce of tenfold vanilla or vanilla oleoresin in the warm solvent. Add sufficient water to make 1 gallon, stir, and pass through a homogenizer just once. Some manufacturers do not include any true vanilla extract or vanilla oleoresin in their formulations.

An example of a kola-flavor emulsion [2] which is suggested for use in the manufacture of beverages sweetened with non-nutritive sweeteners is the following:

KOLA-FLAVOR EMULSION

Lemon oil	⅔ fl oz
Lime oil, fivefold	⅔ fl oz
Cinnamon oil	80 minims
Tannic acid	2.5 oz
Solid kola extract	⅝ oz
Gum arabic	2 oz
Water	1 pt
Caramel, acidproof sufficient for	1 gal

The first three ingredients are mixed together and the oil mixture is rubbed into the gum arabic. About 0.5 pint of the water is added and, after stirring thoroughly, this mixture is passed through a homogenizer or colloid mill to obtain an initial emulsion. The tannic acid and the solid kola extract are mixed with the other 0.5 pint of water and this mixture is added to the initial emulsion of the oils and gum arabic. To this mixture, the caramel is added, slowly with constant stirring. The final mixture is then passed through a homogenizer or colloid mill.

To manufacture the kola-flavoring bottling concentrate, 2.5 fluid ounces of the kola flavor emulsion, 1.25 fluid ounces of 50 per cent citric acid, 15 minims of sirupy phosphoric acid, and 2 ounces of a cyclamate, like Sucaryl calcium, are added to enough water to make 1 gallon of the bottling concentrate. In making the beverage, a throw of 1 ounce of the bottling concentrate per 7-ounce bottle is used, with sufficient carbonated water containing 3.5 to 4 volumes of carbon dioxide. The finished drink contains 0.22 per cent of calcium cyclamate.

In an analogous manner root-beer flavor emulsions can be prepared.[2]

ROOT-BEER FLAVOR EMULSIONS

Ingredient	Sarsaparilla type	Birch-beer type
Methyl salicylate	½ fl oz	⅔ fl oz
Sassafras-oil substitute	⅔ fl oz	⅜ fl oz
Clove oil	⅛ fl oz	⅛ fl oz
Anise oil	⅛ fl oz	⅛ fl oz
Coriander oil		⅛ fl oz
Caramel coloring	5 pt	5 pt
Gum arabic	2 oz	2 oz
Water, sufficient for	1 gal	1 gal

The essential oils and the methyl salicylate are mixed together and the mixture is used to wet the gum arabic by rubbing together. One pint of water is added, the mixture is stirred thoroughly and passed through a homogenizer or colloid mill to obtain the initial flavor emulsion. The caramel coloring is added to the initial emulsion with stirring and then the remainder of the water to make 1 gallon. After mixing thoroughly, the final mixture is passed through a homogenizer or colloid mill.

It is to be noted again, as was pointed out in Chapter 1 and in Chapter 7, that sassafras oil and safrole and other safrole-bearing oils may have an undesirable degree of toxicity. Therefore, it would be preferable to use a substitute containing no safrole in making the preparation or omitting the sassafras oil entirely.

The bottling concentrate is made by adding 5 fluid ounces of sarsaparilla root-beer flavor emulsion and 2 ounces of calcium cyclamate with enough water to make 1 gallon, in the case of the sarsaparilla root-beer bottling concentrate; and 5 fluid ounces of the birch-beer type flavor emulsion with 2.5 ounces of calcium cyclamate with sufficient water to make 1 gallon, in the case of the birch-beer type. With both types, the drink is prepared by using a throw of 1 ounce per 7-ounce bottle with sufficient carbonated water containing 3.0 to 3.5 volumes of carbon dioxide. The sarsaparilla-type drink contains 0.22 per cent and the birch-beer type 0.27 per cent of calcium cyclamate.

4. CLOUDY-BEVERAGE CITRUS-FLAVOR EMULSIONS

Cloudy citrus beverages are very popular. Years ago, it was difficult to make such beverages having a stable cloud. Very often the oil would not remain in suspension after the soft drink was prepared and would form an unsightly ring in the neck of the bottle. In addition, the turbidity would be markedly lessened. The quest for a "ringless" cloudy emulsion flavor was intense. Efforts to vary the usual gum arabic emulsions with gum tragacanth sometimes were successful but no practical solution was found by change of the emulsifying agent alone.

One of the methods for stabilizing an emulsion is to mix two phases, that is the oil phase and the water phase, having equal specific gravities. Such an emulsion will not have to counteract gravity effects in addition to other stresses tending to break the emulsion.

To make use of the stabilizing effect of equal specific gravities of the phases, it is necessary to raise the specific gravity of the flavor oils being emulsified. The citrus oils, which are the ones most important in this category, have specific gravities of the order of 0.85, whereas the final sugar-sweetened soft drinks have specific gravities of the order of 1.03 to 1.06, depending, of course, on the amount of sugar present in the final drink. Artificially sweetened beverages have final specific gravities of the order of 1.0 and those beverages prepared with both sugar and artificial sweeteners have specific gravities in between those mentioned. In all such instances, it is necessary to add sufficient of some material to the citrus oils or other flavor oil to raise the specific gravity of the oil mixture to that of the final beverage. Some 30 years ago, the use of brominated edible vegetable oils was proposed and tried for this purpose.

Brominated oils had been prepared at Abbott Laboratories [3] about that time as possible radiopaque agents for x-ray diagnosis. Brominated olive oil, made by this firm, was tried as a weighting agent for citrus oils and was found suitable for that purpose. The advantages of brominated vegetable oils for the stabilization of cloudy citrus-flavored soft drinks are that they have such relatively high specific gravities that the amounts needed in the flavor-oil mixture are economically sound; they are bland in flavor and odor so that they do not interfere with the citrus flavor; they are miscible with the citrus oils; they are relatively inexpensive; and the Food and Drug Administration has not interposed any objection to their use. In a 1956 report, Dr. A. J. Lehman,[4] Chief of the Pharmacology Division, Food and Drug Administration, United States Department of Health, Education, and Welfare concluded, ". . . there is no reason to believe that brominated olive oil in the amounts now employed in carbonated beverages constitutes a public health hazard."

a. **Physical Properties of Brominated Vegetable Oils**

A number of brominated vegetable oils are being marketed. They are clear, viscous brown liquids. They have a bland taste and slight odor. Brominated vegetable oils are miscible with citrus oils and are insoluble in water.

Brominated olive oil, the first of such oils to be used, has a specific gravity of 1.235 to 1.245 at 25°C. The Abbott Laboratories trademark for this product is Brominol.

Brominated sesame oil has a specific gravity in the range of 1.325 to 1.335 at 25°C. The Abbott Laboratories trademark for this product is Densitol. Formerly they assigned this trademark to brominated apricot-kernel oil.

A light-amber, brominated vegetable salad oil with a fruity odor, having a specific gravity of 1.325 to 1.334 at 20°C, an iodine value of 1 (maximum), and no free bromine, is sold under a trademark of Akwilizer by Swift & Company.

Brominated apricot-kernel oil is an analogous product. One brand had a specific gravity of the order of 1.33. Another brand had a specific gravity of 1.25 to 1.26. Because little apricot-kernel oil is available commercially, the manufacture of the brominated product is seldom carried out on an industrial scale.

Still another brominated oil available commercially is a mixture of brominated olive oil and brominated corn oil. This mixture has a specific gravity ranging from 1.305 to 1.315 at 25°C. It has been marketed under the trademark of Gravinol by the Abbott Laboratories.

A brominated mixture of vegetable oils is made on an industrial scale that has an even higher specific gravity than those mentioned, including brominated sesame oil, namely, 1.330 to 1.340.

b. **Adjustment of Specific Gravity**

The first step in the preparation of ringless cloudy flavor emulsions is the determination of the specific gravity of the citrus oil or citrus oil mixture to be used in the flavor. This can be done by the methods detailed in Chapter 15. This specific gravity can be determined readily by the use of a hydrometer. The specific gravity of the brominated oil must also be ascertained. This is generally given on the bottle or container in which the oil is packaged, but it should be checked from time to time.

The second step is to determine or select the specific gravity of the finished beverage.

The third step is to calculate the required proportions of the citrus oil or citrus oil mixture and the brominated vegetable oil. This may be done easily by the Pearson "square" method.

Procedure—(1) Write the specific gravities of the citrus oil or citrus oil mixture on the upper left hand side of a rectangle as in the illustration. Write the specific gravity of the brominated vegetable oil in the lower left hand side of the rectangle.

(2) At the intersection of the diagonals, write the specific gravity of the final beverage less 0.02.

(3) Subtract the number for the specific gravity of the final beverage from the number for the specific gravity of the brominated vegetable oil and, disregarding the decimal point, write the number at the upper right hand side of the rectangle. Subtract the number for the specific gravity of the citrus oil or citrus oil mixture from the number for the specific gravity of the finished beverage and, disregarding the decimal point, write this number at the bottom right of the rectangle. Using a heavier brominated oil, let us assume that the specific gravity of the citrus oil mixture is 0.84, the specific gravity of brominated sesame oil is 1.34, and the specific gravity of the final sweetened beverage is 1.02. The result will be:

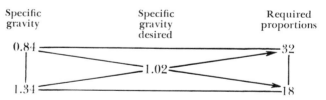

As another illustration, assume a specific gravity of 0.84 for say an orange oil, a specific gravity of 1.24 for a brominated olive oil, and a specific gravity of 1.02 for a sweetened final beverage. The result will be:

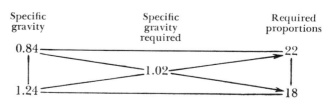

(4) The numbers at the top and bottom of the right side of the rectangle represent the proportions by volume to be used in making the mixed oil, that is the upper right hand number refers to the parts by volume of citrus oil or citrus oil mixture to be used and the bottom right number refers to the parts by volume of the brominated oil to be used.

(5) In some beverage plants, the ingredients are weighed instead of being measured by volume. In this case, it must be remembered that the volume proportions given before must be converted to weight proportions. It must be recalled that density is weight per unit volume, that is, for instance, the weight of 1 milliliter of a given substance is its density at a given temperature. Specific gravity is the ratio of the density of a substance at a given temperature and the density of some substance used as a standard. Ordinarily specific gravity is the weight of a given volume of a substance referred to the weight of an equal volume of water. For the purposes of this calculation, it will be adequate to use the following equation:

$$\text{wt in oz} = \text{vol in fl oz} \times \text{sp gr} \times 1.04$$

The factor 1.04 is required because a weight ounce is not the same as a fluid ounce.

(6) Once the general proportions are obtained, mix the oils and determine the specific gravity by use of a hydrometer or by the methods detailed in Chapter 15. If the desired gravity has not been obtained, add sufficient brominated vegetable oil if the specific gravity is too low, or, conversely, add citrus oil or citrus oil mixture if the specific gravity is too high. The first treatment will increase the over-all specific gravity and the second will decrease it.

It is important to make sure that the oils are mixed thoroughly before taking the specific gravity measurements. All the measurements must be taken at a selected temperature so that the procedure can be repeated exactly when required. The actual specific gravity of the oil mixture is set at 0.02 less than the final beverage specific gravity for it has been found experimentally that this gives maximum stability.

c. **Preparation**

In general, the following procedure [5] can be used to make such cloudy flavor emulsions. Weigh or measure the oils at the selected standard temperature and adjust the specific gravity to the desired value as indicated in the preceding section. Make certain that they are mixed thoroughly and blended. Weigh out the amount of gum arabic or other emulsifying agent and wet it thoroughly with the mixture of oils. Add a given amount of water to the mix and rub all the ingredients together. For formulations in which the final volume is 25 gallons, use 5 gallons of water at this step. After mixing adequately by stirring, run the mixture through a colloid mill or homogenizer to obtain a uniform colloidal dispersion.

Dissolve the coloring matters and the propylene glycol or glycerol in some of the water, if these ingredients are used, add this to the oil-gum-water base mixture, and mix with sufficient water to make the total volume including that of the oil-water mixture prepared equal the set volume. In some of the formulations, this will be 25 gallons. Finally pass the entire mixture through a homogenizer or a colloid mill set so that the particle size will be about 2 microns.

Other general methods of preparation have been suggested. For instance, a known volume of the citrus oil or of the citrus oil mixture and an adequate volume of brominated oil, sufficient to adjust the specific gravity of the mixture, are mixed together thoroughly. Check the specific gravity with a hydrometer and readjust as needed. Place the oil mixture into a vessel and sprinkle the gum on the oil; mix with a paddle or appropriate stirring rod until a uniform mass is obtained, with no lumps. While continuing to stir, add the sirup. This will cause the mixture to thicken. Keep on stirring and add a portion of the water. This will yield an emulsion. Dissolve the coloring matter in the remainder of the water and add this to the oil-gum-sirup-water emulsion; after stirring for an additional 10 minutes, pass through a homogenizer or colloid mill twice. The mixture must be stirred continuously if hand stirring is used. If a high-speed mixer is used for mixing, then only one pass through the homogenizer need be made.

Some processors believe that these emulsions can be made by dissolving the gum arabic in the water, filtering, then adding the oil, and finally the propylene glycol or glycerol, and colors.

Still another variation is to mix the gum with only a portion of the water before adding it to the oil. The rest of the water is used to dissolve the colors and propylene glycol, glycerol, or sodium benzoate. Probably all of these techniques will work and possibly the solution of the gum in the water and filtering has the advantage that all extraneous material in the gum will be removed, but the wetting of the gum by the oil is deemed important and less likely to result in broken emulsions.

d. Formulation

There are three major variations in the formulation of such flavor emulsions. In one of these, only water is used as the solvent; in the second, some sirup is used as an ingredient in making the emulsion; and in the third variation, either glycerol or propylene glycol is used as an ingredient along with water as the solvent. The glycerol and propylene glycol have also some preservative effect.

ORANGE-OIL EMULSION

Orange oil	6.00 l
Brominated vegetable oil, specific gravity 1.33–1.34	3.75 l
Gum arabic	4.50 kg
Water (sufficient for)	100.00 l
FD&C Yellow No. 6	2.25 kg
Sodium benzoate	120.00 g

The orange oil and the brominated vegetable oil are mixed thoroughly together. The specific gravity of the mixture is taken as directed in Chapter 15 or with the aid of a special hydrometer that can be read to the third decimal point. The specific gravity of the mixture would be about 1.025 at 70°F (21°C). If it is not, then the specific gravity must be adjusted by adding brominated vegetable oil or conversely orange oil, depending on whether the specific gravity is less than or more than 1.025. The gum arabic is stirred in and stirring is continued until all the lumps have disappeared. Sufficient water is added to bring the total volume to 100 liters. The mixture is stirred thoroughly and is passed through a homogenizer.

If coloring matter is to be used, it is added and the mixture stirred until the color is dissolved; then it may be passed through the homogenizer again. It is preferable to dissolve the color and sodium benzoate in a portion of the water, adding it before the first homogenizer pass.

A small volume of the orange oil can be replaced by fivefold or terpeneless orange oil to improve the flavor or the flavor can be modified by replacing some of the orange oil with mandarin oil, tangerine oil, lemon oil, or other citrus oils, in the manner described below.

Ringless cloudy citrus emulsion flavors other than orange flavor, and also pineapple flavor, can be made in an analogous fashion, by substituting the oil in question for the orange oil.

ORANGE-OIL EMULSION

Orange oil	1.0 pt
Brominated vegetable oil, specific gravity 1.25	13.0 fl oz
32°Bé sirup	1.0 pt
Gum arabic	12.0 oz
FD&C Yellow No. 6	6.0 oz
Water	13.0 pt

This formulation is an example of the type of emulsion made with a sugar sirup. The ingredients will make 2 gallons of a 2-ounce-to-the-gallon sirup flavor.

The orange oil and the brominated vegetable oil are mixed together thoroughly. The specific gravity is determined as mentioned and adjusted by adding more brominated vegetable oil if the specific gravity of the mixture is too low and more orange oil if the specific gravity is too high. The mixed oils are placed in a suitable vessel and then the gum arabic is sifted in while stirring the mixture. Stirring is continued until all of the gum arabic has been completely dispersed and no lumps of gum remain. While stirring is continued, the sugar sirup is added. This addition will tend to thicken the mixture. Stirring is still continued while 1 pint of water is added. An additional 2 quarts of water are added and stirred into the oil-gum-sirup-water emulsion. The yellow color is dissolved in the remaining gallon of water and, after all of the dye is dissolved, the color solution is added to the emulsion and stirred until the emulsion is uniform. Stirring is continued for a final 10 minutes and then the entire mixture is passed through a homogenizer twice.

A modified orange-flavor emulsion can be made by using 12 ounces of orange oil, 3 ounces of tangerine oil, and 1 ounce of lemon oil in the given formulation.

If the emulsion flavor is not to be used in a relatively short time for the preparation of the actual beverage, that is, if the time of storage exceeds 3 weeks, it has been suggested that 5 grams of sodium benzoate per gallon of emulsion are added. Similar formulations can be made for lemon emulsion, lemon-lime emulsion, and lime emulsion.

LEMON-FLAVOR EMULSION

Lemon oil	12 fl oz
Brominated vegetable oil	10 fl oz
Gum arabic	12 oz
32°Bé sirup	1 pt
FD&C Yellow No. 5	2 oz
Sodium benzoate	10 g
Water, sufficient for	2 gal

The oils in this formulation should be adjusted to a specific gravity of 1.028. This formulation is used in the ratio of 2 ounces of emulsion flavor to 1 gallon of sirup.

LEMON-LIME-FLAVOR EMULSION

Lemon oil	8.0 fl oz
Lime oil	6.0 fl oz
Brominated vegetable oil	12.0 fl oz
Gum arabic	12.0 oz
32°Bé sirup	1.0 pt
FD&C Yellow No. 5	2.0 oz
FD&C Green No. 3	2.2 g
Sodium benzoate	10.0 g
Water, sufficient for	2.0 gal

The oils in this formulation should be adjusted to a specific gravity of 1.028. This formulation is also used in the ratio of 2 ounces of emulsion flavor to 1 gallon of sirup.

LIME-FLAVOR EMULSION

Distilled lime oil	14.0 fl oz
Brominated vegetable oil	12.0 fl oz
32°Bé sirup	1.0 pt
FD&C Green No. 2	2.6 g
Gum arabic	12.0 oz
Sodium benzoate	10.0 g
Water, sufficient for	2.0 gal

The oils in this formulation are also adjusted to a specific gravity of 1.028 and the emulsion is used in the ratio of 2 ounces per gallon of sirup. Formulations which contain propylene glycol or glycerol in addition to water but no sirup have been devised. Abbott Laboratories,[3] pioneers in this field, have suggested certain formulations for the manufacture of cloudy citrus flavors. Among these are the following:

MODIFIED ORANGE-FLAVOR EMULSION

Florida orange oil	3.0 lb
California orange oil	2.0 lb
Tangerine oil	1.0 lb
Mandarin oil	0.5 lb
Orange oil, fivefold	2.0 lb
Orange oil, terpeneless	2.0 oz (avoir.)
Brominated sesame oil	5.0 lb (approx.)
Gum acacia	12.0 lb
FD&C Yellow No. 6	22.0 oz
FD&C Yellow No. 5	5.0 oz
FD&C Red No. 1	9.0 oz
Propylene glycol or glycerol	2.5 gal
Water, sufficient for	25.0 gal

The method of preparation is the same as that previously detailed in the subsection on preparation. The oils are blended, the specific gravity is adjusted at 1.02 to 1.03, the gum is stirred into the oil mixture, 5 gallons of water are added, and after stirring adequately, the mixture is run through a homogenizer. The colors are dissolved in water and the propylene glycol or glycerol is also dissolved in water and both are added to the oil-gum-water emulsion and mixed. After sufficient water is mixed in to make a total volume of 25 gallons, the entire mixture is run through a homogenizer or colloid mill. If brominated olive oil is used instead of brominated sesame oil, about 7 pounds of brominated olive oil will be required.

This modified orange-flavor emulsion is used in a ratio of 2 fluid ounces with either 1.5 or 2.0 fluid ounces of 50 per cent citric acid solution per gallon of 32°Bé sirup. The recommended throw is 1.25 ounces per 7-ounce bottle and 2.25 ounces per 12-ounce bottle, with carbonated water containing 1.5 to 2.0 volumes of carbon dioxide.

LEMON-LIME FLAVOR EMULSION

California lemon oil	4.0 lb
Lemon oil, fivefold	1.0 lb
Distilled lime oil	1.5 lb
Brominated sesame oil	3.1 lb (approx.)
Gum arabic	10.0 lb
FD&C Blue No. 1	0.5 oz
FD&C Yellow No. 5	20.0 oz
Propylene glycol or glycerol	2.5 gal
Water, sufficient for	25.0 gal

The general method of preparation is followed, adjusting the specific gravity to 1.01. This lemon-lime emulsion flavor is used in a ratio of 2 fluid ounces with 3 fluid ounces of 40 per cent citric acid solution per gallon of 30°Bé sirup. The recommended throw is 1.00 ounce per 7-ounce bottle and 1.75 ounces per 12-ounce bottle, with carbonated water containing 4 volumes of carbon dioxide.

If brominated olive oil is used instead of brominated sesame oil, about 4.4 pounds of brominated olive oil will be required.

LEMON-FLAVOR EMULSION

California lemon oil	4.5 lb
Lemon oil, fivefold	1.5 lb
Brominated sesame oil	3.0 lb (approx.)
Gum arabic	10.0 lb
Propylene glycol or glycerol	2.5 gal
Water, sufficient for	25.0 gal

The general method of preparation is followed, adjusting the specific gravity of the oil mixture to 1.01. This lemon-flavor emulsion is used in a ratio of 2 fluid ounces with 2.5 to 3.0 fluid ounces of 50 per cent citric acid solution per gallon of 30°Bé sirup. The recommended throw is 1.00 ounce per 7-ounce bottle and 1.75 ounces per 12-ounce bottle, with carbonated water containing 4 volumes of carbon dioxide.

If brominated olive oil is used instead of brominated sesame oil in this formulation, then approximately 4.25 pounds of brominated olive oil will be required.

Similar formulations can be given for a *lime-flavor emulsion* and a grapefruit-flavor emulsion. To make a lime-flavor emulsion, 5.5

pounds of distilled lime oil are blended with 1 pound of fivefold lime oil and approximately 3.2 pounds of brominated sesame oil or 4.5 pounds of brominated olive oil. The specific gravity is adjusted to 1.01. The flavored sirup is made in the same ratio as that of the lemon-flavor emulsion except that 3.0 to 3.5 fluid ounces of 50 per cent citric acid solution are suggested. The suggested volume of throw and volumes of carbonation are the same as for the lemon-flavor emulsion.

For a *grapefruit-flavor emulsion*, 6.5 pounds of whole grapefruit oil are blended with 1.0 pound of fivefold grapefruit oil and approximately 3.5 pounds of brominated sesame oil or 5.0 pounds of brominated olive oil. The specific gravity of the mixture is adjusted to 1.01. This grapefruit flavor is used in a ratio of 2 fluid ounces of emulsion flavor with 4 fluid ounces of 50 per cent citric acid per gallon of 30°Bé sirup. The recommended throw is 1.00 ounce per 7-ounce bottle and 1.75 ounces per 12-ounce bottle, with carbonated water containing 4 volumes of carbon dioxide.

e. **Sugar-Free Formulations**

Certain modifications must be made in the formulation of flavor emulsions for the manufacture of ringless cloudy sugar-free carbonated beverages. It has been pointed out in a preceding section that the specific gravity of such sugar-free soft drinks is nearly the same as that of water, that is, about 1.0. Since, as mentioned, such emulsions should be prepared with a specific gravity about 0.02 less than the finished beverage, flavor emulsions to be used for sugar-free or "dietetic" soft drinks should be adjusted to a specific gravity of about 0.98. The flavor level should also be increased. Bodying agents, which are discussed in a later section, may be incorporated in the final beverage.

ORANGE-FLAVOR EMULSION FOR SUGAR-FREE SOFT DRINKS

California orange oil	9.0 fl oz
Valencia orange oil	5.0 fl oz
Orange oil, fivefold	4.0 fl oz
Terpeneless orange oil	0.3 fl oz
Brominated sesame oil	7.0 fl oz
Gum arabic	20.0 oz (avoir.)
FD&C Yellow No. 6	4.0 oz
Propylene glycol or glycerol	2.0 pt
Water, sufficient for	2.0 gal

The general method of preparation is followed, adjusting the specific gravity of the mixture of oils at 0.97 to 0.98. This modified orange-flavor emulsion is used in a ratio of 2.5 fluid ounces with 2.0 fluid ounces of 50 per cent citric acid and 2.5 ounces of calcium cyclohexylsulfamate per gallon of water as the bottling concentrate. The recommended throw is 1 ounce per 7-ounce bottle, with carbonated water containing 1.5 to 2.0 volumes of carbon dioxide. The final beverage contains about 0.28 per cent of calcium cyclohexylsulfamate.

LEMON-LIME FLAVOR EMULSION FOR SUGAR-FREE SOFT DRINKS

California lemon oil	9 fl oz
Lemon oil, fivefold	2 fl oz
Distilled lime oil	3 fl oz
Brominated sesame oil	6 fl oz
Gum arabic	20 oz
Propylene glycol or glycerol	2 pt
Water, sufficient for	2 gal

The general method of preparation can be followed, adjusting the specific gravity of the blended oils at 0.96 to 0.97. This lemon-lime flavor emulsion is used in a ratio of 2.5 fluid ounces with 3.0 fluid ounces of 50 per cent citric acid solution and 2.0 ounces of calcium cyclohexylsulfamate in a gallon of water as the bottling concentrate. The recommended throw is 1 ounce to a 7-ounce bottle, with carbonated water containing 3.0 to 3.5 volumes of carbon dioxide. The final soft drink contains 0.22 per cent of calcium cyclohexylsulfamate. Two ounces of lemon-lime color solution can be used to color the bottling concentrate.

5. BODYING AGENTS

In Chapter 2, one of the advantages cited for sugar over artificial sweeteners, discussed in Chapter 3, in carbonated beverages was that sugar gave "body" to the drink. Beverages which are made for those who have to limit their carbohydrate or sugar intake, that is, dietetic beverages may have this lack of body corrected by the incorporation of a substance with either a low calorie value or no caloric or nutritional value. Such substances are known as bodying agents in the carbonated-beverage industry.

Among those which have been used are sorbitol, carboxymethylcellulose, and pectin.

a. **Sorbitol**

Sorbitol, $HOCH_2 \cdot HCOH \cdot HOCH \cdot HCOH \cdot HCOH \cdot CH_2OH$, a hexahydric alcohol, is a dry, white, odorless, crystalline powder in the pure state, melting at 93°C, but the melting point may range from 87° to 97°C. It has a pleasant, sweet cooling taste. Sorbitol is readily soluble in water and is somewhat soluble in alcohol. It may be synthesized from natural sugar or glucose by hydrogenation.

Sorbitol is sold commercially as a solution under several trade names. Arlex is one of these commercial sorbitol solutions. It is an aqueous noncrystallizing sirup consisting principally of sorbitol and some other closely related polyhydric alcohols which prevent it from crystallizing. It is a clear, colorless liquid containing about 83 per cent of polyhydric substances and 16 per cent of water. It has a sweetish, slightly caramel-like odor and a bland, sweet taste. Its average specific gravity is 1.322 and it has an average refractive index of 1.487 at 25°C. Its viscosity, density, and refractive index are higher than those of glycerol and certain glycols under corresponding conditions. This commercial sorbitol solution is completely miscible with water and 92.5 per cent alcohol.

Sorbitol contributes about as many calories to a beverage as does sugar, but since only 1.0 to 1.5 per cent of sorbitol is used in the final beverage in contrast to the 9 to 14 per cent of sugar commonly used, it is clear that the total calorie contribution is much less.

The amount of sorbitol to include in the flavored, artificially sweetened concentrate has been listed in Table 3-3, of Chapter 3. Thus it is suggested that 7 gallons of 70 per cent sorbitol solution are used per 100 gallons of pale dry ginger ale sugar-free bottling concentrate; 8.5 gallons of 70 per cent sorbitol solution per 100 gallons of kola concentrate; and 10 gallons of 70 per cent sorbitol solution per 100 gallons of orange concentrate. Since sorbitol has some sweetening power, it is necessary to reduce the cyclamate a proportionate amount.

b. **Sodium Carboxymethylcellulose**

Sodium carboxymethylcellulose, also known as sodium celluloseglycolate and by a number of trademarks and trade names, such as CMC, Cellofas WFZ, Fondin, sodium Tylose, is a white to cream-colored fine powder or granules or a fluffy hygroscopic solid which is soluble in water at all temperatures, that is, in both hot and cold water in contrast to methylcellulose. CMC is commercially available in different viscosities, ranging from 5 to 2000 centipoises for a 1 per cent solution. It is stable in solutions ranging from pH 2 to 10 and is not affected by the presence of salts. In Chapter 3, it was noted that CMC 70 S High [6] was suggested as a bodying agent for sugar-free soft drinks in the ratio of about 19 ounces per 100 gallons of bottlers concentrate for ginger ales, 27 ounces per 100 gallons of kola-flavored bottlers concentrate, and 34 ounces per 100 gallons of orange-flavored bottlers concentrate. The final beverage will contain about 0.6 to 0.8 per cent of sodium carboxymethylcellulose.

Sodium celluloseglycolate is not attacked by either the acid digestive juices of the stomach or the alkaline digestive juices of the intestines and, therefore, cannot be utilized by the body. Thus it has no nutritive value.

Sodium carboxymethylcellulose requires, as do a number of other bodying agents, good agitation to dissolve it in the concentrate and the beverage.

c. **Other Substances**

In addition to sorbitol and sodium carboxymethylcellulose, other substances have also been suggested as bodying agents. Among these are pectin, demethoxylated pectins, alginates, and other cellulose derivatives. These also have virtually no nutritive value. They are used in concentrations of the order of 0.6 to 0.8 per cent. They require good agitation during addition to the concentrate and the beverage to make certain that they dissolve satisfactorily. The disadvantages of pectin when the beverage is to be canned are noted in the discussion concerning the canning of beverages.

Mention is also made in the literature of the use of Irish moss and of glycerol as bodying agents for carbonated beverages sweetened with non-nutritive sweeteners.

6. FOAMING AGENTS

At times foaming agents are used in carbonated beverages. This practice was probably more common years ago, when root beer and related beverages with high foaming capacity were popular beverages, than it is today. Most carbonated beverages no longer have strong foaming properties and in many beverages, such as fruit-flavored carbonated soft drinks, foaming is considered undesirable. Foaming agents are of three principal types: (1) denatured proteinaceous materials similar to those used in the manufacture of foam-type fire extinguishers but of an edible grade; (2) saponin-free natural extractives like licorice or glycyrrhiza extract obtained from the dried rhizome and roots of *Glycyrrhiza glabra* of different varieties; and (3) saponin-bearing extracts or solutions like those made from soaproot or soapwort, *Saponari officinalis,* Levantine soaproot, *Gysophila strutium,* quillaja or soap bark (quillay bark), *Quillaja saponaria,* and "soap nuts," the fruits of *Sapindus saponaria.*

So far as foaming properties are concerned, particularly when the amount of foaming agent to be used is taken into consideration, the foaming products made with saponins are probably better than those made from denatured proteins and licorice root, but the question of toxicity arises with saponins. Some states forbid the use of saponins or saponin-bearing foaming products in carbonated beverages. The Food and Drug Administration has not ruled against the use of the saponins generally used in soft drinks.

If the use of a foaming agent is deemed desirable for a given product, it is probably best from an economic and production point of view for the beverage manufacturer to obtain the foaming agent from a commercial source rather than to manufacture it himself from the bark or roots of the plant products mentioned before.

Saponin foam extracts usually contain about 10 per cent of saponin and generally 1 ounce is used per gallon of beverage sirup. The final beverage will contain saponin in the order of 0.007 to 0.008 per cent.

LITERATURE CITED

1. M. B. Jacobs, *Synthetic Food Adjuncts,* New York, Van Nostrand, 1947.
2. *"Sucaryl" Sweetened Beverages,* Technical Information No. 42, North Chicago, Abbott Laboratories, 1957.
3. *Brominated Oils,* Technical Information No. 37, North Chicago, Abbott Laboratories, 1957.
4. A. J. Lehman, *Assoc. Food Drug Officials (U. S.) Quart. Bull.,* **20,** No. 2, 71 (Apr. 1956).

5. K. M. Beck, *Food Eng.*, **29**, 139 (1957).
6. G. M. Brenner, U. S. Patent, 2,691,591.

SELECTED BIBLIOGRAPHY

Basic Formulas for Brominated Oils, North Chicago, Abbott Laboratories, 1952.

"How to make Cloudy Ringless Emulsion Concentrates," New York, C. W. Logeman Co.

C. K. Simon, "The Preparation of Ringless Cloudy Orange and other Citrus Emulsions," Long Island City, Dominion Products, Inc.

Brominated Oils, Technical Information No. 37, North Chicago, Abbott Laboratories, 1957.

M. B. Jacobs, ed. *Chemistry and Technology of Food and Food Products,* New York, Interscience Publishers, 1951.

Chapter 9

COLOR AND COLORING

Pleasing color makes a beverage more attractive and desirable. The proper coloring of a beverage is then a matter of importance to every beverage manufacturer for consumer acceptance may depend on it.

1. CLASSIFICATION OF COLORS

The coloring matters used for beverages can be placed into three major categories, namely, (1) natural coloring matters, (2) artificial coloring matters, and (3) synthetic coloring matters. For the purposes of beverage manufacture it is necessary to distinguish between artificial colors, that is, coloring matters that are made by some artifice from natural materials as in the instance of caramel from sugar and colors that are wholly synthetic, such as the coal-tar colors. In the manufacture of nonalcoholic carbonated beverages, artificial and synthetic coloring matters are the most important groups.

Natural colors *per se* are not widely used in the preparation of soft drinks. This is due to several marked advantages that synthetic colors have over natural coloring matters. For instance, coal-tar colors are less prone to spoil and pound for pound are able to color far more of the finished product than a natural color. The coloring power of some synthetic dyes is so great that 1 ounce will easily color over 32,000 ounces of finished product. Indeed, often a concentration of 1 part in 100,000 is used. As long ago as 1888, Sulz,[1] in commenting on the use of rosaniline (not a permitted beverage color) for coloring beverages, pointed out that this dye excelled in brilliancy all vegetable colors (then in use), was cheaper, very intense, and that only trifling amounts were needed.

The synthetic colors can be further classified according to their use in the beverage industry under three headings, (a) primary, (b) secondary, and (c) tertiary colors. These terms are, however, used mainly in describing the FD&C colors to be discussed in a subsequent section.

A *primary* color is one that consists of a straight, unmixed certified FD&C color. Thus, for instance, when amaranth (FD&C Red No. 2) is used alone to provide a ruby red color, as in coloring cherry soda, it is termed a primary color.

A *secondary* color is one which consists of a mixture of two or more primary colors. For example, a certified color mixture for a raspberry red might consist of a mixture of amaranth (FD&C Red No. 2) and ponceau 3R (FD&C Red No. 1), or a green might comprise a mixture of tartrazine (FD&C Yellow No. 5), brilliant blue FCF (FD&C Blue No. 1), and light green SF yellowish (FD&C Green No. 2).

A *tertiary* color is one which consists of a mixture of secondary colors or a mixture of a secondary and a primary color. The color may or may not be mixed with a diluent.

The term *mixture* is applied to a coal-tar color prepared by mixing two or more straight certified colors, or one or more straight colors and one or more diluents.

The word *diluent* designates any component of a coal-tar color mixture except a straight dye. The principal diluents in beverage colors are sugar that is, sucrose, dextrose, lactose, and dextrin.

2. NATURAL COLORS

Very few natural colors are used for the coloring of soft drinks for the reasons noted in a preceding paragraph. Actually with few exceptions, the natural colors used in the manufacture of carbonated and still beverages are those that are present in the true fruit extract used to make and flavor the true fruit beverages. The more important of these are cherry, raspberry, and strawberry. Generally synthetic coloring is added particularly in grape-flavored beverages where a very deep color is desired.

3. CARAMEL

There is only one major artificial coloring matter. This is caramel color. It is defined as the concentrated aqueous solution of the product obtained by heating sugar or glucose until the sweet taste is destroyed and a uniform dark mass results, a small amount of alkali or alkali carbonate being added while heating (*N. F.*). Possibly a broader descriptive definition is that caramel is the mixture of products produced by heating sucrose, glucose, molasses, or other saccharine products within the range of 190° to 220°C. It should be noted that the Food and Drug Administration deems caramel to be an artificial color since it is made by the artifice of partially breaking down sugar.

a. Properties

Caramel is a dark-brown to black, viscous liquid with a characteristic odor of burnt sugar from which property it gets its synonymous names of *burnt sugar* and *burnt sugar coloring*. It has a pleasant, somewhat bitter taste. Caramel (*N. F.* X) should have a specific gravity of 1.30 at 25°C. It is soluble in water and 1 part dissolved in 1000 parts of water forms a solution with a distinct yellowish-orange color. The color of such solutions should not change perceptibly within six hours nor should any precipitate form within this period on exposure to sunlight. Caramel is soluble in water in all proportions and is also soluble in alcohol solutions containing less than 55 per cent by volume of alcohol, but it is insoluble in most of the common organic solvents, such as ethyl ether, petroleum benzine, benzene, chloroform, acetone, and the like.

As a test of its purity, the addition of 0.5 milliliter of phosphoric acid to 20 milliliters of a 1:20 aqueous solution of caramel produces no precipitate. Caramel swells when incinerated forming a cokelike charcoal which burns off only after prolonged heating at high temperature. Not more than 8 per cent of residue remains after ignition.

Caramel is composed of a mixture of substances. When prepared from pure sucrose, caramel is completely soluble in water. When prepared from unrefined sugar, however, certain water-insoluble components are formed. The quantity of these insoluble components increases with increase in the extent of dehydration. The cara-

COLOR AND COLORING

mel made from sucrose is much more soluble in aqueous alcohol solutions than that made from commercial glucose for the latter contains dextrins which, on heating, form alcohol-insoluble derivatives. The addition of alkali carbonate during the heating process apparently increases the coloring power. Caramel acts as a stabilizing colloid in molasses and sirups, but apparently does not affect the viscosity. Nearly all commercial caramels produced in the United States range from 34° to 42°Bé. It is important to use a uniform Baumé of caramel.

Solid and powdered caramels are also available. Solid caramel is generally a reddish-brown, brittle, amorphous, highly deliquescent substance.

b. Manufacture

Caramel can be manufactured, for example, by heating sucrose and other products, such as glucose, molasses, soybean carbohydrate, and other carbohydrate material.

(1) Sucrose

A caramel can be made from sucrose by heating it above its melting point, 186°C but below the point at which it chars. This method is inadequate for commercial production and cannot be relied on to obtain a uniform product.

A relatively simple process is to heat 100 pounds of simple sirup to boiling, add about 150 grams of 28 per cent ammonia water while mixing the batch, and add an aqueous solution of ammonium carbonate containing 250 grams of the compound. Heating is continued until a temperature in the range of 350° to 450°F (176.6°–232.0°C) is obtained.

An alternative method is to heat a solution of sugar in concentrated ammonium hydroxide solution under pressure at 105° to 120°C. Any volatile products formed in the process may be removed subsequently by passing steam through the caramel. Pressure may be released and reapplied in the processing step.

Specific recommendations for the preparation of caramel have been given by Leopold [2] and by Sollich.[3] Leopold found that it is best to heat the sugar at 187°C for 16½ minutes and that longer heating develops a bitter taste. Sollich observed that partial concentration could be done in an open vessel but final concentration was best performed in *vacuo*. Still another recommendation [4] is to heat sucrose at 210°C. It is known that unrefined sucrose caramelizes much more readily than refined sucrose. This is due to the mineral content of the unrefined product.

To prepare a relatively small batch of caramel or sugar coloring,[5] fuse 10 pounds of sugar (sucrose) in a wide, low-form vessel, like an iron frying pan. Heat 5 pints of water to boiling and add it to the fused sugar which caramelizes during the fusion. Boil until a sirup is obtained.

An alternative small-batch process is to stir the fused mass until it is brown, remove it from the heat, and add sufficient water to dissolve the caramelized sugar. A paste may be made by the addition of commercial glucose.

(2) Raw Sugar or Molasses

Caramel can be obtained from raw sugar or molasses by the following procedure. Dilute the raw sugar solution or the molasses and allow to stand until virtually all the suspended particles have settled out. Invert the sucrose with sulfurous acid and remove the excess sulfur dioxide by boiling for half an hour. Neutralize with lime and filter. Boil the filtrate cautiously, using a vessel equipped with a steam coil. Replace the water lost during boiling and continue the process until a product of adequate coloring power is obtained. Evaporate to a viscous consistency.

(3) Starch, Dextrin, and Glucose

Caramel may be prepared from starch by several methods. In Europe, a common procedure is to dampen powdered starch with dilute sulfuric acid. The batch is heated at 100° to 120°C and is held at that temperature until hydrolysis is complete. This may be checked by the addition of 95 per cent ethyl alcohol to a solution of a sample of the batch. A clear solution indicates complete hydrolysis; a turbidity indicates the presence of dextrins. Calcium carbonate is added to neutralize the sulfuric acid. The mixture is allowed to stand; the clear supernatant liquid is decanted from the precipitated calcium sulfate, it is concentrated to 36°Bé, and is filtered. After transfer to a vessel, the hot filtrate is heated to boiling while being stirred. Sodium carbonate is added in the ratio of 3 parts to 100 parts by weight of the batch and after the color deepens, the heat is reduced gradually to avoid carbonization. Partial cooling is accomplished by the addition of hot water. The caramel is extracted with water, filtered, and then diluted to provide several grades of color.

In an alternative procedure [6] of preparing caramel from starch, the starch is saccharified by heating with dilute hydrochloric acid under pressure. The sugar obtained is then heated at 120° to 130°C with a catalyst, such as ammonium carbonate, ammonium sulfate, ammonium chloride, aluminum sulfate, or calcium chloride.

In one process, the caramel is made directly from dextrin [7] without conversion to glucose. The starch is converted in the customary ways to dextrin which is caramelized directly in the presence of ammonia or another alkali.

A caramel paste was made by Di Baja [8] from glucose by hydrolyzing potato and cornstarch in an autoclave with dilute acid. He stressed the importance of quality in the glucose used, recommending 41 per cent as the minimum D-glucose content required in such commercial glucose.

(4) Malt

Malts are specially prepared to provide caramel coloring. Ordinary malt is given an additional processing by steeping, drying, and heating until it is adequately caramelized. The color of these products ranges from yellowish brown to a blackish brown, the second being produced by high-temperature drying. In this instance, the principal sugar being caramelized is maltose.

(5) Lactose

While this sugar has not been used to any great extent for commercial caramel production, it is to be noted that it loses water of hydration at about 130°C and, while melting at 200°C, undergoes caramelization at 160° to 180°C.

(6) Soybean Carbohydrate

Crude stachyose,[9] obtained from the by-products in the extraction of soybean oil by alcohol, can be hydrolyzed with dilute hydrochloric acid; the acid concentration can be reduced to pH 2 to 3 and then the product can be dehydrated by heating at 130°C. While crude stachyose could be used, a commercial caramel could not be prepared by this investigator using direct heating of stachyose.

Instead of using crude stachyose, the lower layer of soybean-oil foots, after treatment with acid, can be used for it is rich in sugars.[9]

This layer is filtered and heated at 130°C. The caramel solution is neutralized with sodium hydroxide to a pH of 6.0 to 7.0. The bitter taste of the caramel preparation was removed by the addition of a small amount of alcohol or by the addition of disodium hydrogen phosphate, monosodium dihydrogen phosphate, or trisodium phosphate.

c. **Use**

Caramel is used extensively as a coloring material, but it must be noted that it has some flavoring properties also. Its principal use by the soft-drink manufacturer is for the coloring of cream soda, root beer, sarsaparilla, birch beer, some ginger ales, and other carbonated beverages. It may also be used by the manufacturer who prepares his own imitation vanilla flavor.

In the last case, the caramel is added in ratios of 7 to 10 parts of caramel to 1000 parts of total ingredients. At other times, the amount of caramel added is dependent on the taste of the processor and the coloring power of the particular caramel batch. Coloring trials, similar to those to be described for use with synthetic coloring matters, may be made to determine the proper amount of caramel coloring matter to use.

4. SYNTHETIC COLORING MATTERS

The synthetic coal-tar colors that are used in the United States with governmental permission are known as certified colors. Often they are termed permitted colors. In order to be used, however, with this governmental permission, they must undergo a procedure known as certification. This consists in submitting to the Food and Drug Administration suitable samples of coloring matter so that the government analysts and scientists may ascertain by chemical, biochemical, toxicological, and medical analysis if the dye is free of deleterious substances. As manufactured for textile or other industrial and commercial purposes, synthetic coal-tar dyes contain impurities some of which may be harmless and others which may be harmful or toxic. These impurities may not detract from the value of a dye for industrial use, but they are highly objectionable in a product prepared for human consumption. It is beyond the scope of this book to discuss this further. The reader is referred to Jacobs [5] for a more complete discussion.

a. Certified Colors

The certified colors permitted to be used in the United States are controlled by the Food and Drug Administration, United States Department of Health, Education, and Welfare. This agency has assigned certain names to the colors which are permitted to be used in beverages and foods. These are known as FD&C colors. In the 1950's, the Food and Drug Administration began an extensive review and an intensive retesting of the toxicology of the certified colors. The status of such FD&C colors that could be used in beverages in 1958 is given in Table 9-1.

b. Pure-Dye Percentage

Pure-dye percentage is defined as the percentage by weight of coal-tar dye in a color. The Food and Drug Administration requires that FD&C Blue No. 1, FD&C Green No. 1, and FD&C Green No. 2 contain at least 82 per cent pure dye in the straight certified color. Most commercial products of this group are prepared to contain 86 to 89 per cent of pure dye in the straight color. All other permitted water-soluble, coal-tar colors are required to contain at least 85 per cent pure dye for certification. Most commercial straight certified dye colors in this group contain from 90 to 92 per cent.

It is very important for the purchaser to note the percentage of straight dye given on the label of any product, for some products, which are ostensibly less expensive than others, actually may cost more in the long run for the amount of available dye because they contain less dye per unit cost.

c. Use

Large quantities of coloring matter are used in the manufacture of carbonated and other nonalcoholic beverages. This has been an accepted practice for many years. Not all of the twelve permitted colors listed in Table 9-1 are suitable for beverage use. Only FD&C Blue No. 1, FD&C Green No. 3, FD&C Red No. 1, FD&C Red No. 2, FD&C Red No. 4, FD&C Yellow No. 5, and FD&C Yellow No. 6 are adequate. The others are unsuitable for various reasons. For instance, FD&C Blue No. 2, FD&C Green No. 1, and FD&C Green No. 2 are not resistant to light; FD&C Red No. 3 is precipitated as the color acid by acids and since most carbonated beverages and soft drinks have some acid, this coloring matter is particularly un-

TABLE 9-1

Recommended Maximum Solutions and Fastness Ratings of FD&C Colors

FD&C	No.	Trade Name	Colour Index No.	Shade	Recommended Maximum Solutions Water, oz/gal		Recommended Maximum Solutions Alcohol, oz/gal, 10%	Relative stabilities* of FD&C Colors to:				
					55°C	70°C	70°C	Sunlight	Oxidation	Reduction	Acid	Alkali
Blue	1	Brilliant Blue FCF	671	Greenish Blue	25	30	30	F	P	P	G	F
Blue	2	Indigotine	1180	Blue	1	1.5	1	P	P	F	P	P
Green	1	Guinea Green B	666	Bluish Green	16	20	16	F	P	P	P	F
Green	2	Light Green SF Yellowish	670	Bluish Green	25	33	25	F	P	P	P	F
Green	3	Fast Green FCF	—	Bluish Green	18	23	18	F	P	P	F	P
Red	1	Ponceau 3R	80	Cherry Red	8	15	8	G	F	P	G	F
Red	2	Amaranth	184	Magenta Red	10	18	8	F	F	P	G	F
Red	3	Erythrosine	773	Bluish Pink	12	16	16	P	F	P	I	G
Red	4	Ponceau SX	—	Light Cherry Red	3	9	5	G	F	P	G	F
Yellow	5	Tartrazine	640	Lemon Yellow	6	18	12	G	F	P	G	F
Yellow	6	Sunset Yellow FCF	—	Yellowish Orange	12	20	15	F	F	P	G	F
Violet	1	Acid Violet 6B	697	Reddish Violet	16	20	16	F	P	P	P	F

* G = Good, F = Fair, P = Poor, I = Insoluble.

COLOR AND COLORING 187

suitable. The seven colors mentioned and FD&C Violet No. 1 and their mixtures are adequate to give almost any color shade desired. See Table 9-2. Only FD&C Yellow No. 5 and FD&C Yellow No. 6 are used as primary colors to any extent, the first for lemon drinks and the second for orange drinks.

(1) Coloring Trials

In order to maintain a uniform color in a beverage, the manufacturer must make coloring trials of every batch of color solution used.

One gallon of water, that is, 3785 milliliters, is transferred to a 4- or 5-liter beaker. Thirty or sixty milliliters of the coloring solution are added and the mixture is stirred thoroughly with a glass rod. This solution now represents the colored stock solution. The standard throw, that is, 37 milliliters ($1\frac{1}{4}$ ounces) of the colored solution is transferred with the aid of a graduated cylinder to an empty soda-pop bottle. Several drops of stock citric acid solution are also added to the bottle and then it is filled to the usual mark with water. The mixture is mixed by inverting the stoppered bottle several times. The contents of the soda-pop bottle are transferred to a Nessler or other comparison tube and the color is compared with that of a standard. The color trial is repeated with more or less color as the case may be. The final trial is checked by making a trial run with soda water.

The 30- or 60-milliliter volume of color solution used for making the simulated sirup is adequate for color solutions containing 4 ounces of dry color per gallon of solution which is customary for red, orange, and dark shades. If a color trial of a light shade is being made, such as lemon, use only a 15-milliliter aliquot.

(b) Preparation of Color Solutions

It is often simplest for the small and medium-sized beverage manufacturer to purchase prepared color solutions from suppliers of such products rather than to make his own solutions. Some manufacturers prefer to prepare their own color solutions. This is because, depending on conditions, it is more economical to do so; it permits greater control; and it ensures secrecy.

When color solutions are made in the plant, several precautions must be observed. The water, as mentioned in a preceding chapter, may contain impurities or hardness which may precipitate the color, or the dye may be affected in other ways so that its coloring power is reduced. Mention has already been made of those dyes that are suitable for beverage colors and their fastness ratings have been tabulated.

(1) Procedure

Weigh as accurately as possible the amount of dry color or colors required or else count out the number of preweighed packets (generally in the form of 1-ounce packets) required. Heat one half to three quarters of the final volume of water to be used to 160°F. Add the dry color, in small portions at a time, until all the color has been added while stirring. Continue stirring until all the dye has dissolved, breaking up any lumps with the stirring rod or paddle. If the color solution is to be used immediately after preparation, add the remainder of the water to complete to volume.

If the color is to be used at some subsequent time, add sufficient cool water to equal four fifths of the final volume, reserving the remaining fifth of the water for the preparation of a preservative solution. Dissolve sufficient sodium benzoate, using 3 to 4 grams per gallon of color solution, in a portion of the remaining water and stir into the color solution. Dissolve citric acid, using 5.5 grams per gallon of color solution, in another portion of the remaining water and add to the color solution. Allow the color mixture to cool to room temperature and make up to the final volume with additional water. Tartaric acid may be used instead of citric acid. Generally 4 ounces of dry color are used for each gallon of solvent, but for light shades 2 ounces of color and for darker shades 8 ounces of dye are used per gallon.

(2) Precautions

Clean Equipment. Since many of the dyes are readily precipitated, reduced or oxidized, it is necessary to use clean equipment in the preparation of color solutions. All cleansing compounds used in the washing of the vessels must be rinsed away thoroughly. For instance, any free chlorine remaining on the equipment after washing will react with the dyes being dissolved and cause fading.

Interference of Metals. Since many of the colors are reduced or precipitated by metals, such as iron, aluminum, zinc, tin, and copper, particularly in acid solution, it is best not to store or prepare color solutions in such metal containers or vessels. For small lots of color solutions, glass vessels such, as are used in a chemical laboratory, are preferred and glass stirring rods should be used. Enamelware, provided it is not chipped, is also suitable, earthenware may be used for the preparation of the same or similar color solutions. For large lots, stainless steel, Monel metal, and noble-metal-lined vessels should be used for the preparation, handling, and storage of color solutions, but even these are not suitable for storage periods exceeding 2 days. Glass-lined vessels and wooden containers are probably best, but care must be exercised to avoid chipping the glass lining. Glass rods or wooden paddles should be used for stirring the mixtures.

Interference of Hard Water. Hard waters contain, as has already been discussed, calcium or magnesium or both metal ions. These metals have a tendency to precipitate the colors by forming insoluble lakes which deposit as a sediment on standing. Such water should be softened by use of water softeners or deionized. Indeed when water softeners are used, care must be observed not to use types that may react with the dyes. It is best to use distilled, deionized, or demineralized water.

Correct Weighing. The utmost care should be observed in weighing out the color to be dissolved, otherwise it will be difficult to obtain the same shade of color with subsequent color batches. This is particularly true when mixtures are prepared. As mentioned, pre-weighed packets obtained from commercial suppliers may be used. It is not adequate to use measuring devices instead of weighing the color because the apparent specific gravity or density of the same dye is very likely to differ from batch to batch. At times the dye may be fluffy; at other times it may be dense. Thus equal volumes will have unequal amounts of pure dye. In addition, the actual percentage of dye should be noted. Twice as much coloring preparation will have to be weighed out if the pure-dye percentage is 45 as will be necessary if the pure dye percentage is 90.

Preservation. In the procedure detailed for the preparation of color solutions, sodium benzoate acidified with citric acid was suggested as a preservative, for beverage coal-tar dyes are attacked by molds and other microorganisms with loss of coloring matter and power. It is necessary to add an acid, such as citric acid, because benzoates act as preservatives only in the presence of acids. The necessity for the addition of an acid accentuates another precaution to be observed. Sodium benzoate, if used, should be added to the color solution first and then the acid should be added. They should not be dissolved and mixed together before addition to the color solution because benzoic acid, which is not very soluble in water, may be precipitated and thus will be more difficult to disperse and dissolve in the color solution.

If it is desired, glycerol, propylene glycol, or alcohol may be used as preservatives in the color solutions. About 25 per cent of the water used should be replaced by these solvents. If alcohol is to be added, it is preferable to cool the color mixture before the addition of the alcohol. This avoids loss of alcohol by evaporation. It should be remembered that all of the water-soluble beverage colors, with the exception of FD&C Red No. 3, are less soluble in alcohol than in water; consequently, if alcohol is to be added, the lowered solubility should be taken into consideration in the amount being weighed out for solution.

It is well to point out again that FD&C Red No. 3, erythrosine, is precipitated by acids; therefore, citric acid and other acids cannot be added to color solutions containing this dye. These color solutions or blends must be preserved by using at least 10 per cent of alcohol.

Solubility. The synthetic coal-tar colors show marked differences in solubility. It is important, therefore, to prevent certain colors, such as FD&C Blue No. 2 and FD&C Red No. 4, which have low solubilities, from being subjected to temperatures below freezing, for the dye may precipitate from solution. Recommended maximum solution concentrations are given in Table 9-1. The addition of a considerable percentage of glycerol and propylene glycol generally prevents this type of precipitation.

The solubility, at any given temperature, of a certified dye is a function not only of the actual solubility of the coloring matter itself, but also of the amount of salt that is present in the original color product. For this reason it is always better to dissolve small portions of the dye to bring it up to color strength rather than to dissolve a large amount of the color. This will avoid any salting-out effect attributable to dissolving the salt with the formation of a salt solution before the dye is dissolved. This is particularly necessary if salt is used as the dye diluent.

Protection against Light. Certain beverage colors, for instance, FD&C Blue No. 2, FD&C Green No. 1, FD&C Green No. 2, are light fugitive, that is, they fade when exposed to the action of light. Most of the colors are affected in some measure by light. Consequently, if they are stored in glass containers, for any length of time, the glass should be of an amber, brown, or green shade; or other means for protection against light should be used. The degree to which a color is fugitive should be taken into consideration in calculating the amount of dye to be used. For in low concentrations, a fugitive color may fade out completely, while in higher concentrations, the fading may not be noticeable for a long time.

Filtration. If the precautions detailed in the previous paragraphs have been observed, it will generally not be necessary to filter the color solutions, for a bright, clear solution of the dye will be obtained. If, however, the solution is cloudy, or it contains dirt, specks, undissolved lumps of color, or sediment, it is necessary to filter the dye solution in order to avoid getting these impurities into the finished beverage.

It is best to allow the color solution to stand for at least 2 days to clarify to some degree by sedimentation. The clear solution can then be filtered by decantation of the supernatant liquid through filter paper. The filtrate should be caught in clean, brown, amber, or green glass bottles. These should be stoppered with plastic caps or closures. If the bottles are equipped with metal closures, they should have nonmetallic interliners. The bottles should be properly labeled and dated to avoid mistakes in use.

The dyes most apt to lump are FD&C Blue No. 1, FD&C Green No. 1, FD&C Green No. 2, FD&C Green No. 3, and FD&C Red No. 1. These dyes should be stirred particularly well when being dissolved.

d. Formulation

In general, coloring solutions can be made by dissolving 4 ounces of a primary, secondary, or tertiary color or dye mixture in water or other solvents, such as alcohol, glycerol, and propylene glycol, to make 1 gallon of solution. Alcohol, glycerol, and propylene glycol have, as mentioned, preservative action.

Coloring trials, performed as detailed in Section 4.c.(1) of this chapter, will assist in evaluating the standard coloring solution for a given beverage. Some representative beverage coloring solution formulations are given in Table 9-2.

TABLE 9-2

Coloring Solutions

Cherry	
FD&C Red No. 1 (Ponceau 3R)	20 oz
Alcohol	30 oz
Water, sufficient for	5 gal
Lemon	
FD&C Yellow No. 5 (Tartrazine)	25 oz
Glycerol	30 oz
Water, sufficient for	5 gal
Lemon-Lime	
FD&C Yellow No. 5 (Tartrazine)	20.0 oz
FD&C Blue No. 1 (Brilliant Blue FCF)	0.5 oz
Propylene glycol	30.0 oz
Water, sufficient for	5.0 gal
Grape	
FD&C Red No. 2 (Amaranth)	15 oz
FD&C Blue No. 2 (Indigotine)	5 oz
Glycerol	30 oz
Water, sufficient for	5 gal
Grape	
FD&C Red No. 2 (Amaranth)	25.0 oz
FD&C Blue No. 1 (Brilliant Blue FCF)	2.5 oz
Propylene glycol	30.0 oz
Water, sufficient for	5.0 gal
Orange	
FD&C Yellow No. 6 (Sunset Yellow FCF)	22 oz
FD&C Yellow No. 5 (Tartrazine)	5 oz
FD&C Red No. 1 (Ponceau 3R)	9 oz
Glycerol	30 oz
Water, sufficient for	5 gal
Orange	
FD&C Yellow No. 5 (Tartrazine)	32.5 oz
FD&C Red No. 1 (Ponceau 3R)	4.8 oz
Propylene glycol	30.0 oz
Water, sufficient for	5.0 gal

Raspberry
FD&C Red No. 2 (Amaranth)	60.0 oz
FD&C Red No. 1 (Ponceau 3R)	3.8 oz
Glycerol	30.0 oz
Water, sufficient for	5.0 gal

Strawberry
FD&C Red No. 2 (Amaranth)	17 oz
FD&C Red No. 1 (Ponceau 3R)	3 oz
Propylene glycol	30 oz
Water, sufficient for	5 gal

Strawberry
FD&C Red No. 2 (Amaranth)	17.5 oz
FD&C Yellow No. 5 (Tartrazine)	2.5 oz
Glycerol	30.0 oz
Water, sufficient for	5.0 gal

LITERATURE CITED

1. C. H. Sulz, *Treatise on Beverages*, New York, Dick & Fitzgerald, 1888.
2. H. Leopold and E. Wagner, *Z. Untersuch. Lebensm.*, **83**, 487 (1942).
3. R. Sollich, British Patent 359,634 (1930).
4. M. B. Jacobs, Ed., *Chemistry and Technology of Food and Food Products*, Vol I, page 93, 2nd Ed., New York, Interscience Publishers, 1951.
5. M. B. Jacobs, *Synthetic Food Adjuncts*, New York, Van Nostrand, 1947.
6. Pei-Sung King, *Ind. Center* (China), **3**, 319 (1934).
7. O. Jungfer, German Patent 582,039 (1933).
8. Atilio Di Baja, *Requind*, **1**, 52, 74 (1932).
9. Y. Iwasa, *J. Agr. Chem. Soc. Japan*, **14**, 1512 (1938); **15**, 473 (1939).

SELECTED BIBLIOGRAPHY

U. S. Food and Drug Administration, *Service and Regulatory Announcements, Food, Drug, and Cosmetic* No. 3, "Coal-Tar Color Regulations," Washington, D. C., 1940; as amended to 1958.

F. Fiene and S. Blumenthal, *Handbook of Food Manufacture*, New York, Chemical Publishing, 1938.

S. Blumenthal, *Food Manufacture*, New York, Chemical Publishing, 1942.

Chapter 10

CARBON DIOXIDE AND CARBONATION

The sparkle of a carbonated soft drink is due to the carbon dioxide it contains. This component adds to the "life" of the beverage and contributes, in some measure, to the tang.

Carbon dioxide is widely distributed in nature, both as the free gas in normal air in which it is present to the extent of 0.03 to 0.04 per cent and as salts and compounds, such as sodium carbonate, sodium bicarbonate, calcium carbonate, and the like. Many rocks are composed in whole or in part of carbonates. Among the more common are marble, limestone, chalk, and magnesite which are principally composed of calcium carbonate and dolomite, a calcium magnesium carbonate.

1. CARBON DIOXIDE

At normal temperatures and pressures, carbon dioxide is a colorless, odorless, noncombustible, slightly acid gas. It does not support combustion. It is heavier than air and has a specific gravity, compared to air, of 1.529. Carbon dioxide condenses to a colorless, refractive liquid at 20°C when subjected to 50 atmospheres of pressure. When this liquid is permitted to evaporate spontaneously, a portion of it is frozen to a white solid. The solid melts at −56.6°C under 5.2 atmospheres of pressure and the liquid boils at −78°C. The solid sublimes forming the gas at normal temperatures and pressures. The liquid has a specific gravity of 1.1 at −37°C. In common with other gases, carbon dioxide is more soluble in water at lower temperatures than at higher temperatures. Thus one volume of carbon dioxide is soluble in 0.6 volume of water at 0°C whereas at 25°C, it requires 13 volumes of water to dissolve 1 volume of carbon dioxide. The solubility of carbon dioxide in sugar solutions is given in Table 10-1.

TABLE 10-1

Solubility of Carbon Dioxide in Sugar Solutions at 60°F and 760 mm Pressure [a]

Sugar, %	Volume of Gas Dissolved
1.0	0.995
2.0	0.989
3.0	0.982
4.0	0.975
5.0	0.967
6.0	0.959
7.0	0.951
8.0	0.943
9.0	0.936
10.0	0.928
11.0	0.918
12.0	0.907
13.0	0.902

[a] After H. W. von Loesecke, *Outlines of Food Technology*, 2nd Ed., New York, Reinhold, 1949.

a. **United States Pharmacopeia Requirements**

Carbon dioxide or carbonic acid gas is described in U.S.P. XV [1] as follows: Carbon dioxide contains not less than 99 per cent by volume of CO_2.

Description—Carbon dioxide is an odorless, colorless gas. Its solutions are acid to litmus. One liter of carbon dioxide at 0°C and at a pressure of 760 millimeters of mercury weighs 1.977 grams.

Solubility—One volume of carbon dioxide dissolves in about 1 volume of water.

Identification

A. Carbon dioxides extinguishes a flame.
B. Carbon dioxide produces a white precipitate when passed

into barium hydroxide test solution, that is, a freshly prepared, saturated solution of barium hydroxide in recently boiled water, the precipitate dissolving in acetic acid with effervescence. U.S.P. XV notes that cylinders of carbon dioxide should be maintained at 25° ± 2°C for at least 6 hours prior to withdrawing samples for the tests and assay, and the results corrected to 25°C and 760 millimeters of mercury.

Acid and sulfur dioxide—Pass 1000 milliliters of carbon dioxide through 50 milliliters of recently boiled water which has been cooled to room temperature. Regulate the flow so as to require 15 minutes for the delivery of 1000 milliliters of the gas. The delivery tube must have an orifice approximately 1 millimeter in diameter and must extend to within 2 millimeters of the bottom of the vessel containing the water. The vessel employed must give a hydrostatic column of 12 to 14 centimeters with 50 milliliters of water. After the passage of the gas, pour the liquid into one of two similar comparator tubes (A) and add 0.1 milliliter of methyl orange indicator solution, prepared by dissolving 100 milligrams of methyl orange in 100 milliliters of water and filtering if necessary. To the other tube (B) containing 50 milliliters of cooled, recently boiled water, add 1 milliliter of 0.01 N hydrochloric acid, and then 0.1 milliliter of methyl orange indicator solution. Viewed downward over a white surface, the liquid in tube A shows no deeper shade of red than in tube B.

Phosphine, hydrogen sulfide, and organic reducing substances—Pass 1000 milliliters of carbon dioxide, under conditions comparable to those in the test for *acid and sulfur dioxide,* through a mixture of 25 milliliters of silver ammonium nitrate test solution, prepared by dissolving 1 gram of silver nitrate in 20 milliliters of water and adding 10 per cent ammonium hydroxide solution, dropwise, with constant stirring, until the precipitate which is first formed is almost, but not entirely, dissolved and then filtering, and three additional drops of 10 per cent ammonium hydroxide solution. No turbidity of darkening should be formed as shown by comparison with another portion of the test solution through which the gas has not been passed.

Carbon monoxide—Collect in suitable flasks 1000 milliliters of carbon dioxide to be tested and, for the blank test, 1000 milliliters of carbon dioxide prepared by treating sodium bicarbonate with hydrochloric acid. Add 10 milliliters of water to 0.5 milliliter of oxalated or defibrinated blood and mix thoroughly. Immediately add 2.5 milliliters of the blood dilution to each flask, stopper, and shake the flasks frequently during 15 minutes. To each flask add 40 milligrams of a mixture of equal parts by weight of pyrogallol and tannic acid. Shake thoroughly, and allow the flasks to stand in the dark for 15 minutes. Pour the contents of each flask into test tubes for observation. The solution from the carbon dioxide being tested shows no pink coloration and matches the gray color produced in the blank test.

Assay—Place a sufficient quantity of mercury in a 100-milliliter nitrometer or gas burette provided with a two-way stopcock and a two-way outlet, and properly connected with a balancing tube. Connect one of the outlet tubes of the nitrometer with a gas pipette of suitable capacity. Place in the pipette about 125 milliliters of 50 per cent potassium hydroxide solution. Draw the liquid (free from air bubbles) through the capillary opening, connection, and stopcock opening in the nitrometer by reducing the pressure in the nitrometer tube and opening the stopcock controlling the connection with the gas pipette. Then close the stopcock. Having completely filled the nitrometer, the other stopcock opening, and other intake tube with mercury, draw into the nitrometer exactly 100 milliliters of carbon dioxide by reducing the pressure in the tube. Close this stopcock. Increase the pressure on the carbon dioxide in the nitrometer tube, and open the stopcock controlling the connection with the gas pipette. Force the entire volume of gas into the pipette. Close the stopcock and rock the pipette gently, providing frequent contact of the liquid and the gas. At the end of 5 minutes, when most of the carbon dioxide will have been absorbed by the liquid, facilitate the absorption of the remainder by drawing some of the liquid into the nitrometer tube and forcing the residual gas back on the surface of the liquid in the gas pipette. Again rock the pipette until no further diminution in the volume of the gas occurs. Draw the residual gas, if any, into the nitrometer tube and measure its volume. Not more than 1 milliliter of gas remains.

b. **Physiological Response**

It is beyond the scope of this book to discuss the physiological response induced by either ingested or inhaled carbon dioxide or carbonic acid gas. It is important to point out that carbon dioxide is not physiologically inert as is so often mistakenly stated. Carbon dioxide is a respiratory stimulant. It is the regulator of the breathing function. An increase in the amount of carbon dioxide breathed results in an increased rate of breathing. In some instances it is necessary to induce this effect.

Persons who work with carbon dioxide must understand that very high concentrations of carbon dioxide in the air are dangerous. The maximum allowable concentration for an 8-hour working day has been set at 5,000 ppm, that is, 0.5 per cent by volume. Concentrations of over 15 per cent by volume in the air may cause death. Care must be exercised when working with carbon dioxide.

c. **Preparation**

Years ago considerable amounts of the carbon dioxide used for carbonating beverages were generated directly on the premises of the bottling plant by treating some carbonates such as limestone, chalk, or sodium bicarbonate, or another carbonate mentioned in a previous section, with sulfuric and hydrochloric acids. It should be mentioned, however, that a greater yield is obtained by the use of sodium bicarbonate than the others mentioned and the yield from this source does not vary as it may when chalk or limestone is used as the raw material. Since liquid carbon dioxide and solid carbon dioxide became available on a commercial scale, it is no longer necessary to prepare carbon dioxide on the premises.

In general, the carbon dioxide used by the carbonated beverage industry is obtained (1) as a by-product of some industrial process, as for instance in the manufacture of lime from limestone; (2) by the burning of coke, coal, natural gas, or other fuel and the recovery of the carbon dioxide formed; or (3) as a by-product in the fermentation and brewing industries. In the last case, the carbon dioxide given off during the action of the yeast on sugar is recovered. This carbon dioxide has, however, if it is not additionally purified, a fermentation odor, which, while not objectionable if the carbon dioxide is to be used to carbonate beer, renders it not particularly suitable for soft drinks.

Considerable quantities of solid carbon dioxide, known commonly as Dry Ice are used for carbonating. The relative merits of using solid carbon dioxide or liquid carbon dioxide are still matters of discussion.

2. CARBON DIOXIDE VOLUME

The volume of carbon dioxide in the finished beverage is a most important factor, for it is the quantity of carbon dioxide dissolved in the beverage that gives it its sparkle and governs the length of time, along with other factors, such as temperature, that the beverage will continue to effervesce.

Henry's law states that the amount of a gas dissolved by a given volume of a solvent at constant temperature is directly proportional to the pressure of the gas with which it is in equilibrium. According to this law, the amount of carbon dioxide dissolved by water at a given temperature is proportional to the pressure of the carbon dioxide on the water. This law is, however, conditioned by the nature of the molecule as it exists in the gaseous state and as it exists in solution. In the instance of carbon dioxide, as far as carbonated drinks are concerned, variations from Henry's law are not large.

At atmospheric pressure, the amount of carbon dioxide dissolved by water will depend solely on the temperature. It has already been pointed out that this solubility is greater at lower temperatures than at higher temperatures. The unit of measurement that has been adopted by the beverage and bottling industry as standard is the *volume*. This is defined as the amount of gas in milliliters that a given volume of water will absorb at atmospheric pressure, namely 760 millimeters of mercury and 60°F (15.5°C). These are arbitrary points set by agreement. This condition registers as zero on the scale of gages commonly used to measure the volumes of carbon dioxide absorbed in carbonated beverages.

Thus at 60°F beverage water will absorb 1 volume of carbon dioxide, represented as zero on carbon dioxide volume gages. When the pressure is increased to 15 pounds (actually 14.7 pounds), that is, the pressure of one additional atmosphere, the water will absorb 2 volumes of the gas and for each additional 15 pounds or atmospheres of pressure an additional volume of carbon dioxide will be absorbed. Reduction of the temperature will, as mentioned, permit the water to dissolve greater amounts of carbon dioxide. When the temperature is reduced to 32°F (0°C), 1.7 volumes of carbon dioxide will be absorbed and for each additional increase of 15 pounds pressure or each additional atmosphere there will consequently be an additional absorption of 1.7 volumes of carbon dioxide. Thus, if a bottle was filled with carbonated water at 32°F under 30 pounds of pressure, the volume of carbon dioxide dissolved in the water will be:

$$1.7 \times 3 = 5.1 \text{ volumes}$$

For intermediate pressures and temperatures, intermediate volumes of the gas are absorbed.

3. DETERMINATION OF VOLUME

To estimate the volumes of carbon dioxide in a bottle of a carbonated drink, it is necessary to know the temperature of the beverage and the pressure of the contents of the bottle. The data are determined with gages and thermometers.

In the customary method, a tester consisting principally of a gage having a hollow spike with holes in its sides is used. The bottle is inserted from the side into the slot provided in the neck of the carbon dioxide tester and is secured in place by tightening with a threaded stem. The pressure gage is inserted until the needle point touches the crown or stopper. There is a sniff valve on the gage stem which is kept closed until the needle point of the pressure gage is forced through the crown or cork. The reading is noted on the gage. The sniff valve is opened and the pressure allowed to escape until the first bubble rises in the bottle contents. The sniff valve is now shut and the bottle is shaken until the gage has reached the maximum pressure. The gage is read again. It may be lower than the first reading which was taken before sniffing the head pressure in the bottle, although it is usually about the same as the final pressure when the sample is cold, that is, at 40 to 50°F. The gas-volume tester and crown are removed from the bottle and a thermometer is used to read the temperature of the liquid contents. The exact details of the method are given in Chapter 15.

4. CARBONATION

Since carbon dioxide is more soluble in cold than in warmer water, it is advantageous to cool the water before sending it to the carbonator. Water coolers are commercially available that can deliver a constant supply of water at about 34°F sufficient for carbonators with capacities of as much as 3600 gallons per hour.

In one type of equipment, water from the beverage manufacturer's supply is fed to a distributing pan over the top of stainless-steel plates. It then runs down both sides of these cold plates, losing heat to the plates and being chilled in turn. The cooled water is then collected in a reservoir below the plates. The water flow is cut off automatically when the water reaches a high level. As water is drawn off for the manufacture of the beverages and the level drops in the reservoir, the refrigerating cycle begins over again.

The use of cooled water gives the bottler several marked advantages. These are more uniform operation, the proper bottling conditions uninfluenced by weather, reduction of boiling at the filler, less wear on the equipment, a probable saving of carbon dioxide gas since there will be less loss, the probability of fewer stops and delays, and reduction in costs.

The commercial bottler usually uses liquid carbon dioxide kept in heavy steel tanks which contain 50 pounds of the gas under sufficient pressure to keep it liquid. The tank is connected to a type of equipment known as a carbonator by means of a metal connecting tube. The carbonator is a device by means of which a large surface of water can be exposed to carbon dioxide gas under pressure. The combination of large surface and pressure enables the cooled water to absorb the gas quickly. The number of volumes of carbon dioxide absorbed depends on the factors discussed in the section on gas volume.

Commercial carbonators vary in capacity from 250 to 3600 gallons per hour. It would not be feasible to describe all of such equipment. They are of two major types, one in which the water is only carbonated, Figure 6, and the other in which the water is cooled and carbonated, Figure 7.

a. **Carbonator**

One of the first type of these devices known as the "Cem Saturator" works on the simple principle mentioned previously. It contains no agitating devices, such as paddles. In this device, carbon dioxide gas enters through a gas connection and permeates the tank at operating pressure. Cold water from the water refrigerator is pumped into the tank through the water inlet, is forced up a hollow center pipe column, and is ejected through a specially designed nozzle as a smooth flowing sheet of water. The water is spread out-

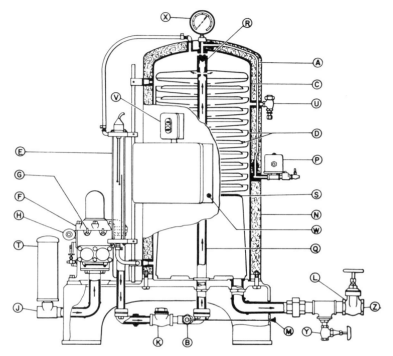

FIGURE 6. Diagrammatic Sectional View of a Typical Cem Saturator Model C (Courtesy of the Crown Cork & Seal Co., Inc., Philadelphia, Pa.)

A—Tank insulation
B—Safety valve
C—Stainless-steel absorption tank
D—Film plates
E—Electrode water level-control system
F—Water pump
G—Vent valves (4)
H—Bypass valve for siphon filling

J—Water inlet
K—Water check valve
L—Carbonated-water outlet valve
M—Safety-valve knob
N—Tank cover
P—Automatic-snift solenoid valve
Q—Center pipe
R—Water spray nozzle

S—Antisplash shell assembly
T—Air chamber
U—Gas-inlet check valve
V—On-off switch
W—Flood switch
X—Gas-pressure gage
Y—Drain valve
Z—Carbonated-water outlet

ward from the nozzle so that it flows over the underside of an inverted film plate. It then flows smoothly downward toward the center column over the film plate immediately below. This action is repeated continuously, the water at all times flowing as a thin film from plate section to plate section until it runs off the skirt of the lowest film plate into the carbonated-water storage area at the bottom of the carbonator tank. Carbonation is continuous and progressive and is governed by regulating the pressure of the gas at the drum for the amount of carbonation desired or required. From the carbonator, the carbonated water flows to the filler.

The quantity of carbonated water stored in the bottom section of the carbonator tank is regulated automatically by adjustable electrode-type probes which actuate and turn off the pump motor. The pump operates only when the level of the carbonated water falls to an established low level and stops operating when the level of the water reaches the high water-level probe. The high level can be set as desired for maximum operating efficiency by adjusting the electrode housing.

Any air which may enter the saturation tank in the water is automatically vented by an automatic snift. This device can be used with all carbonators. The automatic snift assembly functions to release air each time the pump motor starts.

It is imperative that carbonators be made of corrosion-resistant metals such as stainless-steel plates, bronze cylinders, connecting rods and pistons and stainless-steel piston rods. The equipment must also be of sanitary construction so that no added sanitary problem results from its use.

b. **Carbonator and Cooler**

Representative of the type of carbonator that cools and carbonates at the same time is the "Carbo-Cooler" of Figure 7. In this type of equipment, warm, uncarbonated water enters at the top going into a distribution pan from which it flows downward over stainless-steel cooling plates becoming carbonated with the carbon dioxide being admitted from the side. The cooled, carbonated water flows to a reservoir from which it can be conducted to the filler, at about 34°F.

A modification of this equipment, shown in Figure 8, serves as a premix device (see Chapter 11). Water and sirup, measured by a type of proportionating meter known as a "Syncrometer" enter the top head of the Carbo-Cooler. The sirup flows into a stainless-steel distributing trough where it is partially diluted by the incoming water. The diluted sirup flows down part of the stainless-steel cooling surface being cooled and partially carbonated. The water flows down the balance of the cooling surfaces and is cooled to

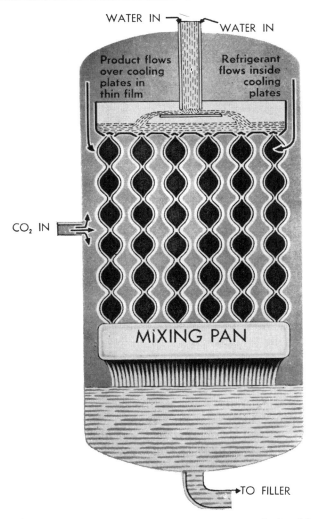

FIGURE 7. Mojonnier Carbo-Cooler (Courtesy of the Mojonnier Bros. Co., Chicago, Ill.)

about 34°F and carbonated. Both water and sirup flow to a mixing chamber in which they are thoroughly mixed and from there go to the reservoir. The finished beverage flows to the filler at about 36°F and the bottles or cans into which it is filled require no additional mixing.

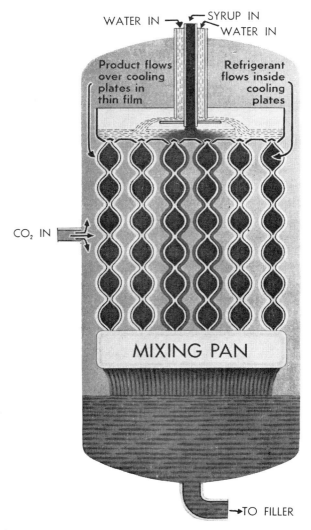

FIGURE 8. Mojonnier Premix Carbo-Cooler (Courtesy of the Mojonnier Bros. Co., Chicago, Ill.)

In both of these methods the cooling is accomplished in an atmosphere of carbon dioxide, so that no additional air is absorbed by the water or by the mixture of sirup and water while it is being cooled and carbonated.

c. Direct-Service Carbonation

In more recent years there has been an expansion of carbonated beverage dispensing from the soda-fountain counter to the vending machine that dispenses drinks in a cup as contrasted to those that dispense drinks in sealed containers. For the latter, the carbonated beverage is prepared in the customary manner at the bottling manufacturer's plant, then placed in dispensers that can be actuated by a deposited coin with subsequent release of the bottled drink. In the open-cup dispenser, the final drink is made by the dispenser itself. The coin actuates the machine which consecutively drops a cup or container into position below an outlet port, pumps prepared sirup and carbonated water to a mixing chamber where the drink is mixed, and then releases the mixed drink to the container. In such dispensing equipment water is carbonated as needed.

Generally these fountain and vending machine carbonators are relatively light in weight, of the order of 50 to 60 pounds, and have capacities of 25 to 100 gallons per hour. The water, usually just filtered city water, is sprayed into a mixing chamber by a pump and mixed with carbon dioxide gas under pressure. The carbonated water is held in a stainless-steel tank and the motor of the pump is actuated and shut off by probes such as stainless-steel, silver-tipped electrodes when the level falls below the low-level point, or rises to the high-level point. The soda-water outlet is connected to the soda-water faucet or to the sirup-mixing chamber in the vending machine.

5. FORMULATION

Some beverages taste better with high carbonation, for example, ginger ales. Others taste better with low carbonation, for instance, orange flavors. The numbers of volumes of carbon dioxide usually recommended for various carbonated soft drinks are listed in Table 10-2. Club soda, sparkling water, seltzer, Vichy, and similar sparkling waters are carbonated at high pressure to contain 4 to 5 volumes of carbon dioxide. The carbon dioxide content of some commercial beverages is given in Table 10-3. See also Chapter 12.

TABLE 10-2

Carbonation of Soft Drinks

Flavor	Gas, volumes
Birch beer	3
Celery	2.5–3.0
Cherry	3.0–3.5
Cream	2.5–3.5
Ginger ale	
golden	4.0–4.5
pale dry	4.0–4.5
Grape	2.0–2.5
Grapefruit	3.0–3.5
Kola	3.5–3.7
Lemon	3.5–4.0
Lemon and lime	3.4–3.7
Lime	3.5–4.0
Orange	0.9–2.0
Pineapple	2.5–3.0
Raspberry	3.0–3.5
Root beer	3.0–3.5
Sarsaparilla	3.0–3.5
Sparkling waters (Club soda)	4.5–5.0
Strawberry	3.0–3.5

TABLE 10-3

Carbon Dioxide Content of Commercial Carbonated Beverages [a]

Name of product	Pressure, pounds per sq in	Temperature	Volume of CO_2 in 1 volume
Coca Cola (many samples)	26–30	45	3.6–4.0
National beer	20	45	3.1
Carbonated waters	28	40	4.2
Canada Dry ginger ale	30	50	3.6
Canada Dry water	26	50	3.3
Seven Up	32	50	3.7
Crown Cola	34	50	3.9
Pepsi Cola	34	50	3.9
Rock Creek ginger ale	60	50	6.0

[a] P. Valaer, *Champagne*. Alcohol Tax Unit, Bur. Internal Revenue, 1948.

TABLE 10-4

Increase of Pressures with Increase of Temperatures in Carbonated Beverages (Warth) [a]

Temperature °F	Pressure in pounds per sq in. developed with volumes of carbon dioxide					
	3.0	4.0	5.0	6.0	7.0	8.0
40	16	26	36	46	56	66
50	23	40	46	59	72	84
60	30	44	59	74	86	103
70	37	56	72	86	107	124*
80	46	66	86	106	126*	146*
90	56	79	102	125*	148*	172*
100	65	91	118*	144*	170*	197*
110	75	104	137*	163*	193*	223*

* Pressures at which there is danger of crown leakage.
[a] After H. W. von Loesecke, *Outlines of Food Technology*, 2nd Ed., New York, Reinhold, 1949.

TABLE 10-5

Permissible Carbonation, Twelve-Ounce Cans [a]

Maximum temperature °F	Carbonation permitted, approximate maximum	
	Standard ends, gas volumes	Heavier ends, gas volumes
110	3.70	4.00
120	3.40	3.70
130	3.15	3.45
140	2.90	3.20

[a] After, "Technical Aspects of Cans and The Canning of Carbonated Beverages," New York, American Can Company, 1953.

6. PRESSURE–TEMPERATURE RELATIONSHIPS

It has already been pointed out that as the temperature goes up, there is an increase in pressure in a container of a carbonated beverage. In the case of bottled carbonated beverages, there may be leakage past the crown as the pressure goes up. These relationships are shown in Table 10-4. In the case of cans, increase in pressure beyond permissible limits may cause the can to be strained or distorted. This hazard is increased if the manufacturer wishes to use an in-can process of pasteurization. The permissible temperatures corresponding to volumes of carbon dioxide are given in Table 10-5 and the pressure-temperature relationships for 12-ounce cans are given in Table 10-6.

TABLE 10-6

Pressure in Relation to Beverage Temperature in Twelve-Ounce Miracans [a]

Temperature of beverage	50°F	70°F	90°F	110°F	120°F	130°F	140°F
Carbonation, volumes measured at 50°F			Pounds per square inch				
2.0	11	19	29	39	43	48	52
2.5	17	28	39	51	57	63	69
3.0	24	37	50	64	71	79	86
3.5	30	46	61	77	85	93	103
4.0	36	54	72	90	99	108	120
4.5	42	61	82	103	112	123	137

[a] After, "Technical Aspects of Cans and The Canning of Carbonated Beverages," New York, American Can Company, 1953.

7. LOSS OF CARBONATION

The loss of carbonation at the filler or after the bottle has been opened is a serious problem in the manufacture and sale of carbonated beverages. Such sudden or rapid loss of carbonation is termed *boiling* when it occurs at the filler and *gushing* when it occurs as the bottle cap is removed. Mention has already been made of the fact that cooling the water and the sirup prior to carbonation prevents boiling at the filler.

The rapid loss of carbon dioxide or gushing of carbonated soft drinks has been discussed in detail by Insalata.[2] It is due to a number of causes, the principal ones being (1) presence of discharge nuclei, (2) excess air in the drink, (3) unnecessary agitation, and (4) improper storage.

Nuclei in the beverage act as discharge points for carbon dioxide gas. Therefore, as soon as the cap is removed and pressure is released, there is an almost instantaneous release of carbon dioxide with a resultant gushing action. Not all nuclei have this property, for size and shape of the particles appear to have importance. Thus rough, irregular-shaped particles are much more likely to act as nuclei for the release of carbon dioxide than smooth, round particles.

While dirty bottles are the principal source of such disturbing nuclei, dusty and moldy crowns, protozoa and algae, and oil droplets are other common sources. The difficulties in the case of dirty bottles may result from poor soaker-compartment control in the bottle washer, with consequent improper cleansing of the bottle interiors; or from carry-over from the last compartment of the soaker to the rinse-water section; or from loose brush bristles.

Insalata points out that soaker carry-over may contain ferric oxide, aluminum hydroxide, and compounds of iron, nickel, cobalt, copper, tin, and other metals. He stresses that as little as 1 ppm of these metals has been implicated in causing gushing. Proper soaker operation and maintenance will overcome and avoid this source of trouble.

Wildness may result from loose bristles of brushes themselves or from particles attached to such bristles. It has been suggested that black nylon brush bristles do not accumulate deposits of particles, are easier to cleanse and keep in sanitary condition, and are more readily detected in the bottles as they pass the inspection light. The use of all-nylon brushes has greatly diminished the severity of this problem.

Scratches on the interior bottle walls provide places which are equivalent to nuclei in causing release of carbon dioxide. Such scratches result from improperly adjusted bottle soaker brushes that can scratch the bottoms of the bottles and from worn brushes that can scratch the sides if the metal of the brush comes in contact with the bottle. To avoid such difficulties the brushes in such soakers must be inspected carefully periodically and replaced when necessary.

Since moldy and dusty crowns can contribute particles which will cause gushing, care must be taken to store these items in such a manner that they will not become moldy or dusty. They should be stored in a cool, dry area of the plant away from direct contact with the warehouse floor. The cartons containing the crowns should be wiped free from dust before they are opened. In addition, the hopper for crown delivery should be equipped with an air line to blow away any dust in the crowns, and the hopper, itself, should be covered.

In Chapter 5, the need for preparation and use of water with special requirements for bottlers' purposes was stressed and detailed. The removal of organic matter, including algae and protozoa, was considered. If a water supply is used that contains organisms of this nature, they may reproduce and form a flocculent precipitate some of which settles out and some of which may remain suspended. These flocs form excellent nuclei for the release of carbon dioxide and in addition detract from the appearance of the beverage.

Because these organisms get into the bottling system by way of a water supply, when provision has not been made for the preparation of an independent water supply, the entire system may become contaminated, as well as the final product. Even the sand and carbon filters may become contaminated. Insalata suggests in such an instance that a reverse flush, using 10 ppm of chlorine and starting from the bottom of the carbon filter be employed to correct this difficulty. This treatment will destroy the organisms in the water, cleanse the sand and carbon filters, disinfecting them at the same time, remove pockets of algae that may be in the carbon filter, and assist in preventing channeling of the carbon.

Oil droplets may be carried over into the carbonated water with the carbon dioxide gas formed when solid carbon dioxide is converted to gas. Such oil particles can also cause wildness and gushing. Not only do such oil droplets cause the rapid release of carbon dioxide, but they may form unsightly rings in the neck of the bottled beverage thus causing a poor appearance. They also reduce the surface tension at the air-liquid interface. To overcome and avoid this source of gushing, it is best to clean the solid carbon dioxide converter at least every week.

Excessive amounts of air in a carbonated beverage may also cause gushing. Air is less soluble in water than carbon dioxide. Consequently, there will be a greater tendency for air to escape once the bottle is opened than for carbon dioxide. As the bubbles of air rise, however, they induce what may be considered a chain reaction and initiate the escape of carbon dioxide also.

The air present in a carbonated beverage gets into it from the water and the sirup. Not all of the air may be displaced during the filling operations and some air may get into the bottles as they travel from the filler to the crowner. Several precautions will reduce this source of difficulty. The water used may be deaerated if desired, the carbonator should be equipped with a snift valve, and the fillers should also be equipped with snift valves.

The reasons why excessive agitation just prior to opening or just after opening a bottle result in gushing are not entirely clear. It may be that such agitation brings the carbon dioxide into contact with the nuclei which cause release of the gas. This is a factor that is not in direct control of the beverage manufacturer, but cautioning the consumer is advisable.

A few additional causes of gushing should be mentioned. These are freezing in storage, stratification of the contents, and long storage. The last two causes can be controlled by rotation of the bottles during storage and the first by keeping the temperature above freezing.

LITERATURE CITED

1. *The Pharmacopeia of the United States of America* (The United States Pharmacopeia) *Fifteenth Revision*, U.S.P. XV, U. S. Pharmacopeial Convention Inc., Washington; Easton, Pa., Mack Publishing, Dec. 1955.
2. N. F. Insalata, *Food Eng.,* **29**, No. 4, 83 (1957).

SELECTED BIBLIOGRAPHY

H. W. von Loesecke, *Outlines of Food Technology,* 2nd Ed., New York, Reinhold, 1949.

"Technical Aspects of Cans and the Canning of Carbonated Beverages," New York, American Can Co., 1953.

A. W. Noling, Ed., *Soft Drinks in Cans* and *Supplement to Soft Drinks in Cans,* Hurty-Peck & Co., Indianapolis, 1954.

M. B. Jacobs, Ed., *Chemistry and Technology of Food and Food Products,* Vol. III, Chapter L, New York, Interscience Publishers, 1951.

H. E. Medbury, *The Manufacture of Bottled Carbonated Beverages,* Washington, Am. Bottlers Carbonated Beverages, 1945.

H. E. Korab, *Technical Problems of Bottled Carbonated Beverage Manufacture,* Washington, Am. Bottlers Carbonated Beverages, 1950.

"Mojonnier Carbo-Cooler," Bull. No. 333-4, Chicago, Mojonnier Bros., 1955.

Chapter 11

BOTTLING AND CANNING

Every preliminary step in the manufacture of a soft drink, namely, the purification of the water, preparation of sirup, preparation of flavoring, preparation of coloring, acidulation, and carbonation, has been considered in detail. All of these must now be put together to make the beverage.

Following Toulouse,[1] we can set up a flow diagram (Figure 9) for the manufacture of carbonated beverages. This comprises three major subdivisions: (1) the raw material section in which we have (a) raw water, sugar, carbon dioxide, acids, flavors, and colors, (b) alkali and detergents, and (c) bottles, cans, crowns, and cases; (2) the manufacturing section in which (a) the flavored and acidulated sirup and carbonated water are prepared, or in which the flavored and acidulated sirup mixed with water and subsequently carbonated is prepared, (b) the bottles are washed or the cans are rinsed, and (c) the bottles and cans are filled, crowned or sealed, and labeled; and (3) the filled storage and shipping sections.

1. CONTAINERS

Until the 1940's, virtually all carbonated beverages were packed in glass. As mentioned in Chapter 1, one of the most significant developments in the soft drink industry was the machine made bottle. This development permitted the manufacture of bottles of such uniformity that all steps, such as washing, filling, crowning,

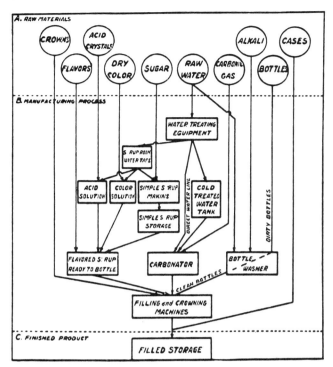

FIGURE 9. Flow Sheet for Carbonated Beverage Manufacture [J. H. Toulouse, *Ind. Eng. Chem.*, **26**, 765 (1934)]

and labeling could be done by machine. In the 1950's, the use of cans for carbonated beverages made substantial progress.

a. **Bottles**

In the United States a variety of shapes in bottles is used by the carbonated-beverage industry, but the capacities of these bottles have been relatively standardized. Carbonated-beverage bottles usually contain $6\frac{1}{2}$, 7, 8, 10, 12, 26, 28, or 32 fluid ounces. The 16-ounce bottle may become more popular. Most bottles used are of the refillable type for this permits greater economy, but in the 1950's there was some attempt to utilize throw-away glass bottles also.

The versatility in glass manufacture has permitted the creation of many privately designed and distinctive bottles to which color may be applied for "permanent" labeling and for specialty decoration.

All bottles for carbonated beverages, whether new or used, must be thoroughly washed immediately before filling. This topic will be discussed in some detail in a subsequent chapter. Very likely the greatest advantage of glass bottling is economy. Undoubtedly another important factor, particularly since most carbonated beverages for home use are purchased by women, is the visual appeal of the beverage and the esthetic appeal of the product in glass. Another very important advantage of the glass bottle is its relative freedom from corrosion. The advantages and disadvantages of bottles used for carbonated beverages may also be inferred from the discussion of the advantages and disadvantages of cans for carbonated beverages.

Pickering[2] points out that the life of a carbonated-beverage bottle is 4 to 6 years and that in that time it is used from thirty to thirty-five times. There is a replacement rate of bottles of the order of 3 to 4 per cent per annum of which 0.50 to 0.75 per cent is due to in-plant bottle breakage. Pickering stressed that several steps could be taken to increase the life of such bottles. Among these were the installation of automatic casing and uncasing equipment, palletization, great care in the use of bottle-handling equipment, care by personnel in the handling of bottles, use of full-depth paper six-bottle carriers is preferable, use of proper dividers in cases, adequate provision for the sorting of flavors, and use of good-quality crown pullers which is essential on vending machines.

b. Cans

Because the use of cans in the manufacture of carbonated soft drinks is a relatively newer development, somewhat greater consideration will be given to them. This does not mean that the author favors the use of cans over bottles. Each type of container has its advantages and disadvantages. Indeed, the one single advantage of economy of glass over cans may be, to any individual beverage bottler, the overriding factor involved in the choice of the one container over the other.

There are two principal types of carbonated-beverage cans. These are the crown-sealed can and the flat-top can. Flat-top cans are manufactured in three sizes: $6\frac{1}{2}$-ounce cans ($2\frac{1}{8} \times 4\frac{1}{8}$), 10-ounce cans ($2\frac{15}{32} \times 4\frac{13}{16}$), and 12-ounce cans ($2\frac{11}{16} \times 4\frac{13}{16}$). The flat-top cans have so-called hard or inert liners, whereas the crown-top cans have heavy wax linings.

Carbonated beverages may be considered as corrosive products. For this reason, they have to be packaged in cans capable of withstanding acid attack, just as other acid foods, such as fruit and fruit products, sauerkraut, and pickles. Carbonated beverages are more difficult to package in cans than other types of acid foods because (a) the pressure exerted by the carbon dioxide requires a can of much greater strength; (b) the large variety of carbonated beverages and the differences in formulation in any given flavor from one manufacturer to another results in a large variety of responses to the can; (c) the flavor, odor, and appearance of carbonated soft drinks are sensitive to small amounts of dissolved metals.

(1) Fabrication

In the fabrication of such cans, corrosion resistant tinplate of very carefully controlled chemical composition is required. The flat sheets of tinplate are first coated with a baked-on epon-type enamel. This enamel is heat resistant and permits the side seam to be soldered without any effect on the inside coating. After the can bodies are formed and soldered, a spray coating is applied to the inside seam area. Subsequently the cans are sent to spraying machines where the entire inside of the cans is sprayed with a vinyl-type coating, followed by heating in an oven to dry the top coating and drive off any residual solvent. The ends also receive a two-coat enamel covering, but both coats are applied to the flat sheets before fabrication.

This treatment gives the entire inside at least two coats of enamel, with the seam receiving three coats. This lining is used for all beverage flavors. It appears entirely inert and does not absorb the flavoring from the beverage nor give the beverage any taste.

Carbonated-beverage cans are made with a tabbed side seam to give the container sufficient strength to hold highly carbonated beverages. As mentioned in the chapter on carbonation, lighter-end 12-ounce cans can hold up to 3.7 volumes and heavier-end 12-ounce cans can hold up to 4.0 volumes.

The outside of the cans must be protected against rusting and this is done by lithographing the can with a label. A protective coat of varnish is put on the bodies and ends and the outside of the can also gets a side-seam stripe after the can is soldered.

(2) Corrosion

Despite the fabrication treatment, corrosion of cans may occur at minute discontinuities or fractures in the coating. A manufacturer of such cans describes two types of internal corrosion of cans. One is a surface corrosion which proceeds gradually and tends to spread laterally rather than to penetrate the metal at a localized point. This type of corrosion presents no perforation hazard and can generally be controlled within acceptable quality limits by good canning, storage, and distribution practices.

The other customary corrosion is the pitting type which attacks at localized areas and tends to perforate the metal with very little visual evidence and without noticeable effect on product quality. This type of corrosion is largely prevented by use of steel-base plate of very closely controlled chemical composition.

Oxygen of the air, both inside and outside the can, will cause corrosion of both the inside and outside of the can. It is clear that the elimination of air from the inside of cans containing canned beverages is far more important than in bottled beverages. The rule of thumb for air in canned beverages is less than 4 milliliters per 12 fluid ounces.

It can be pointed out that many carbonated beverages have no tendency to cause pitting. A few formulations, however, have been found unsuitable for canning for they have this perforating tendency to such a degree that they make the shelf life of such beverages uneconomical.

A slight amount of corrosion of the inside of a can is considered normal. On the basis of their tendency to corrode the cans, soft drinks can be classified into a number of groups: (a) mildly corrosive, namely, the kolas and root beers; (b) mildly to moderately corrosive, namely, ginger ales and lemon-lime beverages; (c) corrosive, namely, fruit-flavored drinks as a class; and (d) corrosive and sensitive to small amounts of dissolved metal, namely, carbonated water.

As certain substances are corrosion accelerators, every effort must be made to keep these substances out of the beverage if cans are to be used. Trace amounts of copper will promote container corrosion. If the beverage manufacturer has copper-bearing equipment it may be best not even to consider canning.

Sulfur compounds are also corrosion accelerators. One can manufacturer mentions that sulfur dioxide is sometimes used as a bleaching agent in sugar and other ingredients of light-colored beverage sirups and advises that no beverage which is known to contain sulfur dioxide or reduced forms of sulfur, either present as a residue or purposely added, should be canned unless test packs prove canning to be feasible. In this connection, it may well be pointed out that a number of organic sulfides are present in minute amounts in natural flavors and in fortified flavors as, for instance, methyl beta-methylthiolpropionate in pineapple flavor, methyl mercaptan in coffee flavor, and dimethyl sulfide in a number of fruit flavors. This may explain why fruit flavors, as a class, are so corrosive.

As pectins and demethoxylated pectins, particularly low methoxyl pectins that are sometimes used as bodying agents in sugar-free and low-sugar carbonated beverages, have corrosive activity, such products also should be test packed.

While phosphoric acid, which, as discussed in Chapter 4, is used as the acidulant in kolas and some other nonfruit-flavored carbonated beverages, generally requires no special precautions as far as canning is concerned, when it is used in conjunction with drinks that contain certain dyes, the beverage may be affected and should, therefore, be tested.

It is good manufacturing practice to test any new formulation of a carbonated beverage by test packing. This is an essential step in beverages that are to be canned and it is a wise precaution to consult the can manufacturer if a change in the formulation is to be made.

A number of substances have been tried as corrosion inhibitors. Of these ascorbic acid, that is, vitamin C, has been found to be most suitable for carbonated beverages. Ascorbic acid also has antioxidant action and thus its use may be economically sound in beverage manufacture.

(3) Advantages and Disadvantages

The advantages claimed for the use of cans for carbonated beverages fall into two groups: consumers' advantages and manufacturers' and distributors' advantages. In so far as the consumer is concerned: No deposit is required and after use, the container can be discarded thus eliminating the storage of cans as contrasted with the storage of bottles which must be returned; cans cool more quickly than glass bottles and can be more easily stored in a refrigerator; cans require less storage space in the kitchen, pantry, etc.; cans weigh less than bottles and, therefore, are easier to handle; and there is no danger from bottle failures or cuts from chipped bottle necks.

From the point of view of the manufacturer and the retailer, the advantages of cans are that no money is tied up to repay deposits for bottles and cases; there is no need for bottle sorting; no space is required for bottle storage; no losses result from broken bottles; and cans take up less space on the retailer's shelves than do bottles.

Some of the disadvantages have been indicated in the discussion concerning the fabrication and corrosion aspects of cans. Canned beverages cost more than bottled beverages in reuseable bottles. Despite artistic lithography, canned beverages do not have the visual appeal of bottled beverages. The return of a deposit may be a nuisance to a retailer, but it is a joy to some youngsters and to the consistent buyer the initial deposit has long been forgotten in the turnover. The shelf life of canned carbonated beverages is less than that of bottled carbonated beverages and, of even greater importance, the retailers' profit is greater with returnable bottles.

2. CLOSURES

It was stressed in Chapter 1 that the invention of the crown cap by William Painter in 1892, in the United States, was a development of the first magnitude in the bottled-beverage industry. Hundreds of different devices and means were used for the closing of bottles before the invention of the crown cap. Its simplicity of construction should not, however, mislead one, for the present crown cap is a product of exhaustive research and manufacturing efficiency. The metal for such crowns must be of uniform gage and ductility, of even temper, and it must be resistant to corrosive attack both from the beverage it seals in the bottle and its outside environment.

Originally, natural cork was used for the crown inserts but over the past two decades the use of composition cork has replaced the natural product. The ground cork is mixed with appropriate binders to yield a more uniform product. Paper-thin spots of impervious materials like aluminum or vinyl plastics are used along with the cork inserts. The fact that cork can be compressed without permanent distortion and flowing, that is, its resilience, is the property that makes cork a highly effective pressure seal.

In the 1940's and 1950's, considerable work was done to develop a plastic liner which would have good flavor characteristics, inertness to the packaged product, resilience, and high sealing effect. Sufficient progress had been made along these lines for the introduction of a new crown in 1956. This crown was the first major change in crowns in about 60 years. It has a slightly shorter metal skirt to accommodate the thinner plastic liner with a concentric ring pattern.

Such liners are made from domestic materials consisting of vinyl resins, inert fillers, and a plasticizer. In contrast, cork must be imported. The claimed advantages for plastic liners over cork inserts are that the plastic liner does not pick up moisture as does cork; spot inserts are eliminated; since plastic liners are thinner less storage space is required for them; and they leave cleaner bottle lips. Another and an important advantage of the new crown with the shorter skirt is that since the skirt is shorter less storage space is required for the finished crown also.

3. BOTTLE FILLING

Clean bottles, prepared as detailed in Chapter 13, are inspected as they leave the washer and before they go automatically through the consecutive steps of filling, crowning, mixing, and labeling. All defective and cracked bottles are removed by an inspector. Filling is customarily carried out by two methods commonly termed the three-stage process and the premix process.

a. Three-Stage Process

The filling method most commonly used in the carbonated-beverage industry is the three-stage process. In this method, a measured amount of the flavored and acidified sirup is first placed into each freshly washed and inspected bottle as it passes through that part of the filling unit termed the siruper (Figure 10). The bottles are then carried along to the second section of this machine where carbonated water from the carbonator is added. In order to maintain the proper carbon dioxide content in the beverage, the filler is constructed in such a manner that a constant counterpressure is maintained during the filling cycle, thus preventing any carbon dioxide from escaping. In a third section of the machine, the crowns described in the preceding section are dropped from a hopper and are automatically pressed on the bottles, sealing them hermetically.

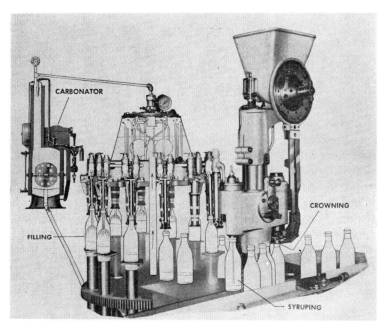

FIGURE 10. Diagrammatic View of a Bottle-Filling Unit (Courtesy of the Crown Cork & Seal Co., Inc., Philadelphia, Pa.)

This unit is a twelve-spout Red Diamond filler—formerly manufactured by the Liquid Carbonic Corporation, Chicago, Ill. The patent rights were purchased by the Crown Cork & Seal Co., Inc., Philadelphia, Pa., a number of years ago and it is now known as the "Cem 1-12."

In one type of bottle filler, that can fill 375 bottles per minute, the bottles are conveyed to the filler by means of a flat-top chain conveyor. The bottles have their first contact with the filler at the timing screw. This device spaces the bottles evenly on the live conveyor chain and accelerates the speed of the bottles to syncronize with the speed of the filler. After the bottle is locked against the rubber centering cup at the bottom of the liquid valve, the mechanical latch trips the outside valve operating lever opening the charging valve. This establishes the gas counterpressure in the bottles. Chipped, cracked, or broken bottles cannot establish such counterpressure and so the gas charging valve closes automatically. Bursting bottles automatically close the liquid and charging valves. Both of these precautions prevent loss of liquid and carbon dioxide.

The filled and capped bottles from the three-stage filling process move on to another automatic machine by means of which the contents of the bottles are mixed thoroughly by a combined whirling and end-over-end motion. Finally the bottles pass in front of a powerful light for inspection. Occasionally a magnifying glass is used as an aid in observation. If the bottles do not have a permanent label they move on to a labeling machine.

b. **Premix Process**

In the premix process of filling, the proper volumes of sirup and water required for each bottle are automatically measured by a continuous metering system. These volumes are mixed, cooled, and carbonated simultaneously by being forced through a special type of carbonator. The carbonated beverage obtained in this manner is piped directly to a filler machine by which it is fed into bottles which are than carried to a crowning section of the machine to be capped.

One of these systems is known as the "Syn-Cro-Mix" system. The unit shown in Figure 11 comprises separate water and sirup pumps mounted on a base plate and separate water and sirup meters connected to a differential valve for controlling the rate of flow of the liquids at a constant predetermined proportion.

The water and sirup pumps of a sanitary type are connected by a V-belt drive. Deaerated water is sent over to the unit, termed a Syncrometer, and is pumped at a constant rate of flow. Sirup is pumped from its mixing tank, or from the sirup tanks or drums, to its independent meter. The two meters are of the piston type mounted at the bottom on a tie bracket for rigidity and have a brass plate at the top on which the differential valve is mounted. A cut brass pinion is mounted on the drive shaft of each meter whose size determines the ratio of mix to be used in the drink. For instance, if 1 ounce of sirup is to be mixed with 5 ounces of water, that is a 5 to 1 ratio is required, the size of the gear is determined by this ratio.

The driven gears of the differential valve are between the two meters and operate on a hollow threaded spindle supporting them. The two differential gears are directly above each other. The low gear operates from the water meter and when this meter stops, the

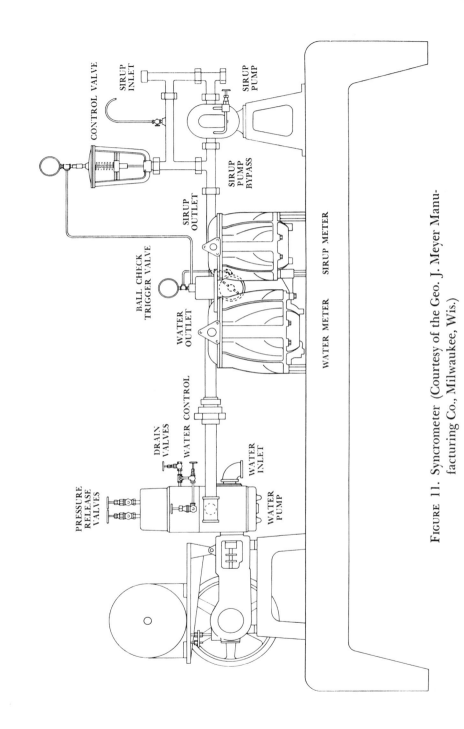

FIGURE 11. Syncrometer (Courtesy of the Geo. J. Meyer Manufacturing Co., Milwaukee, Wis.)

gear also stops. The sirup-meter gear, however, continues to turn on the hollow threaded spindle and raises itself against the solenoid needle valve shutting off the unit. If the sirup meter stops, the process is reversed and the water-meter gear lowers itself to actuate the solenoid. The Syncro lever valve extends from this shaft and operates against an air-driven diaphragm that fluctuates constantly, permitting any excess sirup to be by-passed around the sirup pump. There is always an excess of sirup pumped but the differential valve permits only the exact amount of sirup to go into the final mix at which the gears are set. Thus the flow of water and sirup are designed to be in exact proportion. If the water or sirup supply fails at any time, the unit stops automatically.

The meters are not connected mechanically and thus are free to pass either water or sirup without restraint. Leakproof packing units are used instead of packing glands. The meters do not require adjustment by an operator once they are set. The Baumé of the sirup must, however, be kept constant to insure a proper ratio in the mix and to maintain the uniformity of the finished product. The Syncrometer is intermittent in its operation for it starts and stops on demand of the carbonator. Different ratio gears are available for different mixes required for different flavors, including beverages containing fruit pulp and of low carbonation.

In the premix system, since the beverage has been completely mixed, it is unnecessary for the bottle to go through a mixing step and it passes directly to the inspection section. If the bottle does not have a permanent label it then moves on to the labeler.

Premix filling systems are relatively newer methods of filling than the three-stage process.

In the premix system, a filler similar to that described in the preceding section, is used. The mixed carbonated water and sirup flow from the filler through the liquid valve into the neck and down the sides of the bottle. Pressure in the unfilled neck portion of a bottle is slowly sniffed through a sniff orifice. The filled bottles pass to an 8- or 10-head crowner which is connected to and is an integral part of the filler. The timing of both is syncronized.

4. CAN FILLING

The over-all processes of filling cans with carbonated beverages are entirely analogous to bottle filling, but there are sufficient differences to warrant description. Cans, coming from the manufacturer, must be rinsed to free them of any extraneous matter, such as fragments of fiber or other loose material. The cans may be automatically discharged from their packing cases to the rinse line, but more often they are dumped by hand from cases to an unscrambler from which they pass on automatic runways to the rinser where they are rinsed with either fresh cold water or warm water, as preferred by the bottler. The cans are then carried to the filler in an inverted position so that they can drain.

a. **Premix Process**

In the customary premix system, treated and filtered water is passed through a deaerator and is then sent through a proportionator which automatically measures both water and sirup into a mixing, cooling, and carbonating unit. The finished carbonated beverage is fed automatically to the filler as needed. The filled cans go to closing machines which can handle 200 to 500 cans per minute. Such closing machines must be equipped for gas-flow closure so that the air from the headspace of the container can be displaced with carbon dioxide. If this is done efficiently, only a few milliliters of air will remain in the can. See Figures 12a and 12b.

The filled and closed cans must be warmed to room temperature and preferably higher to aid drying and to prevent sweating after casing, for cans that are cased while they are wet, or which sweat after they are cased, are more subject to external rusting and may damage the cases. Such warming may be accomplished by warm-water sprays. This serves the double purpose of washing away any of the beverage that may have spilled on the outside of the cans. If a washing temperature of 100°F is adopted, the washing will also serve as a check against defective containers reaching a consumer, for this temperature is about the highest to which the cans will be subjected in subsequent handling. A can dryer may be included in the line, but other arrangements, such as compressed-air jets, can also be employed to blow away the water from the tops and sides as the cans pass through the runway.

The can line should be equipped with an automatic weigher or an x-ray level detector to reject underfilled cans. The level detector can be used to record the number of overfilled cans also. The x-ray level detector has the advantage that it does not depend on the weight of the fill. This is important where beverages of different

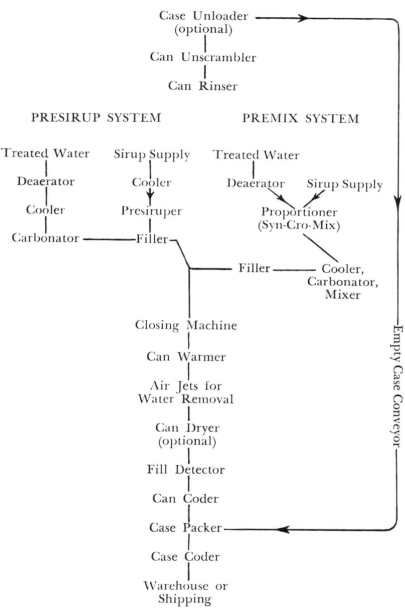

FIGURE 12a. Flow Sheet of Carbonated-Beverage Canning Line

Baumé are being processed, for such beverages would have different specific gravities and thus the same volume would have different weights. If an automatic weigher is used to control underfill, the weight adjustment would have to be readjusted each time a different specific gravity beverage is being filled into the cans. This is avoided by the use of the x-ray device.

Since prolonged heat treatment is inadvisable, the cans should be stored, as soon as possible, in well-ventilated stacks to cool.

It is advantageous to date code each can so that it can be identified should any question concerning it arise. It is also advantageous to code the cases to facilitate rotating the stocks of beverages. Codes may also be used to identify each batch. It is customary to place the code on the bottom end of each can. The final step is to send the cans to the case packer, case sealer, and from there to the shipping floor to warehouse.

b. **Three-Stage Process**

A three-stage system or presiruping method can also be employed with cans. In this instance, a measured amount of flavored and acidified sirup is first placed into each freshly rinsed can which is then carried to a filler, usually low-pressure filling machines. The filler adds the required amount of carbonated water and then the can passes to the closing machine. When the three-stage system is used, the cans must, as with bottle filling, be rotated in some manner to mix the contents. This can be accomplished by having the cans go through twists in the runway. In other respects the filling and sealing methods are the same as those in the premix system.

While it is not customary to include a deaerator in the three-stage process, it is best to have a deaerator or some other method of purging the air from the water.

5. BOTTLING FRUIT BEVERAGES

There was an important difference in the bottling technique of carbonated beverages which contained fruit pulp or other cells of fruit. During the bottling of such products, the sirup had to be stirred constantly so that an even distribution of the pulp could be maintained. This stirring had to be performed in such a manner that no air was incorporated into the finished beverage for as Toulouse [1] pointed out years ago, not only does the air cause foam-

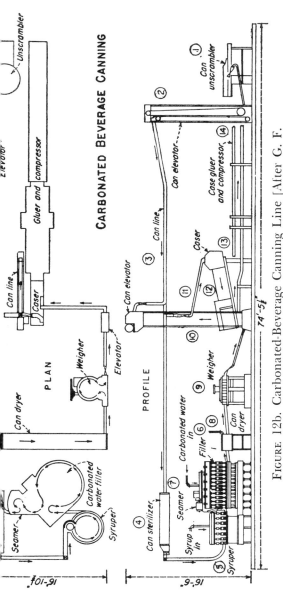

FIGURE 12b. Carbonated-Beverage Canning Line [After G. F. Kay, *Food Eng.*, **26**, No. 2, 86 (1954)]. Copyright *Food Engineering*, New York, N.Y.

In this compact beverage-canning line (16 feet 6 inches by 75 feet 5½ inches), the cans are dumped on an unscrambler (1), elevated (2) to an overhead can line (3), then sterilized (4). The product is filled in a synchronized siruper (5), and carbonated-water filler (6), then the cans are seamed (7). The product temperature is now raised to 72°F in the can dryer (8). Next, each can is weighed (9), then elevated (10), and rolled through twists and turns to mix the product (11). Casing and gluing steps follow (12), (13), and (14). The plan drawing (top) further elucidates the layout.

ing at the filler, as stressed in the chapter on carbon dioxide and carbonation but the oxygen in the air may injure the flavor of the fruit beverage.

In more modern technique, the fruit juice or the fruit flavor may be homogenized so that the particle size of the pulp is reduced to a very small value and no separate stirring may be required. Bottling machinery is available that will handle both ordinary sirups and pulp-bearing sirups. See Chapter 12, section 3.

It must be remembered, however, that additional care must be taken of the bottle-filling equipment when pulp-bearing beverages are made, for the pulp, even when homogenized, may cause clogging of the valves and may get caught in parts of the machinery which will then become difficult to clean and sanitize.

LITERATURE CITED

1. J. H. Toulouse, *Ind. Eng. Chem.,* **26**, 765 (1934).
2. L. K. Pickering, *Natl. Bottlers' Gaz.,* **76**, No. 888, 22 (1957).

SELECTED BIBLIOGRAPHY

Plant Operation Manual, Am. Bottlers Carbonated Beverages, Washington, 1940.
H. E. Medbury, *The Manufacture of Bottled Carbonated Beverages,* Washington, Am. Bottlers Carbonated Beverages, 1945.
M. B. Jacobs, Ed., *Chemistry and Technology of Food and Food Products,* Vol. III, Chapter L, New York, Interscience Publishers, 1951.
"Technical Aspects of Cans and the Canning of Carbonated Beverages," New York, American Can Co., 1953.
G. F. Kay, *Food Eng.,* **26**, No. 2, 86 (1954).
R. Kühles, *Handbuch der Mineralwasser-Industrie,* Lübeck, Antäus-Verlag, 1947.
R. H. Morgan, *Beverage Manufacture,* London, Attwood & Co., 1938.
"Meyer Dumore Syn-Cro-Mix Beverage Filling System," Bull. No. SY. 126-2, Milwaukee, Geo. J. Meyer Manufacturing Co., 1952.

Chapter 12

COMPOSITION OF CARBONATED BEVERAGES

In the foregoing chapters each of the ingredients used for the manufacture of carbonated soft drinks was discussed in detail and the percentages of ingredients used in beverages have been tabulated. In the preceding chapter the actual processes used in manufacturing have been detailed and discussed. In this chapter the composition of carbonated beverages as finished products will be treated.

1. SPARKLING WATER, SODA WATER, AND SELTZER

Most consumers of carbonated beverages make little or no distinction between the terms, sparkling water, soda water, carbonated water, club soda, Vichy, seltzer, and table water (Tafelwasser). While no sharp line of differentiation need be made in most instances, some discussion of the terms is pertinent.

a. Sparkling Waters

Sparkling water may be considered the general term for all carbonated beverages consisting of carbonated water with or without added salts, the principal salts used being sodium carbonate, sodium bicarbonate, sodium chloride, and sodium sulfate. When salts are used, as discussed in a succeeding subsection, a concentrated solution of the salts may be made as the analogue of a sirup and the bottling and filling may proceed along the lines already detailed, with the concentrated salt solution being used in place of the sirup. Thus the siruper will add the predetermined throw of salt solution to the bottle, which is subsequently filled with carbonated water or the concentrated salt solution is fed through a proportionator to the premixer and then carbonated.

Virtually all sparkling waters are carbonated at high pressure, namely, 4 to 5 volumes of carbon dioxide corresponding to 44 to 60 pounds per square inch with the lower pressures at 60 to 75°F.

The term, sparkling, refers to the property of substantial effervescence and was probably adapted from its use to characterize wines as sparkling wines, such as champagne, a white sparkling wine, or sparkling Burgundy, a red sparkling wine, in contrast to still wines, such as port or sherry.

The composition of some sparkling waters is given in Table 12-1.

TABLE 12-1

Composition of Sparkling Waters [a]

Type	Character	Total solids ppm	Carbonation, gas volumes	pH
Sparkling water	mildly saline	250–1000	4–5	4.5
Club soda	mildly saline	1000–2000	4–5	5.5
Vichy	very saline	2000–3000	4–5	5.0

[a] Modified from M. Sivetz, in *Bottlers' and Glass Packers' Handbook*, p. 29, New York, McGraw-Hill.

b. Carbonated Water

The term, carbonated water, should be limited to the product comprising carbonated distilled water, carbonated deionized water, or carbonated treated water. While the term, sparkling water, is appropriate for such products, the designation of such products as soda water is incorrect for it implies that such water contains soda or salts. In England carbonated waters are termed aerated waters.

c. Soda Water

As explained in Chapter 1, the term, soda water, is a generic term being a common synonym for all carbonated soft drinks. More specifically the term, soda water, means carbonated water containing soda. Thus in Great Britain, soda water must contain a minimum of 5 grains of sodium bicarbonate, or roughly ⅓ of a gram, per pint (British). The composition of soda waters varies considerably, depending in some measure on the hardness of the water used for the manufacture of the beverage and on the particular recipe the manufacturer favors. In general, sodium carbonate, or sodium bicarbonate, and sodium chloride are the principal components. The composition of some European types of soda water are given in Table 12-2.

TABLE 12-2

Soda Water Formulations [a]

Salt	grams per 26.5 gallons								
	1	2	3	4	5	6	7	8	9
Sodium carbonate, crystallized	250	200	180	150	75	—	—	—	—
Sodium bicarbonate	—	—	—	—	—	200	150	100	480
Sodium chloride	—	25	20	30	20	60	50	130	30
Sodium sulfate, crystallized	—	—	—	—	—	15	10	—	—
Calcium chloride	—	25	—	30	20	—	—	—	10

[a] After Rudolf Kühles, *Handbuch der Mineralwasser-Industrie*, 4th Ed., Lübeck, Antäus-Verlag, 1947.

In manufacturing soda water, the hardness of the water used must be taken into account. Thus if a deionized, distilled, or soft water is used, a representative formulation would contain 14 ounces and 6 drams (approximately 400 grams) of sodium carbonate, 4 ounces of calcium chloride, and 5.5 ounces of magnesium chloride per 100 gallons, whereas with hard water, a representative formulation would contain 1 pound 9 ounces of sodium carbonate and 5.5 ounces of sodium chloride per 100 gallons.

d. **Seltzer**

The comments of the preceding subsections apply to seltzer or seltzer waters. Originally such carbonated waters were designed to imitate the natural mineral waters. Consequently the number of salts used in the preparation of seltzer waters is greater than that of soda water. A simple formulation contains 6.5 ounces of sodium chloride, 9.0 ounces of sodium bicarbonate, and 0.5 ounce of sodium sulfate. Three ounces of this mixture are used to make 100 gallons of the finished seltzer. Typical compositions of seltzer of European origin are given in Table 12-3. A comparison of Tables 12-2 and 12-3 shows that seltzer generally contains considerably more common salt, sodium chloride, than does soda water.

TABLE 12-3

Seltzer Formulations [a]

Salt	grams per 26.5 gallons										Seltzer spring
	1	2	3	4	5	6	7	8	9	10	
Sodium carbonate	300	200	160	100	100	80	75	32	—	—	132.07
Sodium chloride	150	100	100	50	125	70	25	163	168	100	160.11
Potassium chloride	—	—	—	—	3	—	—	—	—	—	4.66
Sodium sulfate	—	50	10	36	—	—	50	10	—	—	5.17
Potassium sulfate	—	—	—	—	—	—	—	—	—	10	29.45
Magnesium chloride	—	—	—	—	10	5	—	—	—	10	27.04
Calcium chloride	—	—	25	—	5	10	10	—	—	—	
Sodium phosphate	—	—	—	—	—	—	—	—	8	10	0.13
Sodium bicarbonate	—	—	—	—	—	—	—	—	210	160	

[a] After Rudolf Kühles, *Handbuch der Mineralwasser-Industrie*, 4th Ed., Lübeck, Antäus-Verlag, 1947.

2. SPECIALTY BEVERAGES

In Chapters 7 and 8, the preparation and utilization of specialty flavors, such as the kolas, ginger ales, sarsaparillas, and the like, were treated in detail. The composition of commercial beverages of this type are given in Table 12-4.

a. Kola Beverages

A full discussion of the group of kola-flavored carbonated beverages was given in Chapter 7. The general composition of such carbonated beverages is given in Table 12-4. Toulouse [1] conducted a series of investigations of kola carbonated beverages in 1932 and 1933. He found the average sugar concentration to be of the order of 10.68 per cent and the acidity to be 44 grains per gallon. Noling [2] found that the average composition of thirty samples of kola carbonated drinks in cans was 10.1 per cent sugar, pH 2.6, and gas volume 2.8; the range in these samples was 8.0 to 12.5 per cent sugar, pH 2.3 to 3.3, and carbonation from 1.5 to 3.6. Beattie [3] suggests that a kola should have 9.5 to 10.0 per cent of sugar, an acidity in the range of 2.8 to 3.1, for below 2.8 the beverage is too sharp and above 3.1 it is too flat, and a carbonation of 3.3 to 3.6 volumes, for carbonations below 3.3 are too flat.

TABLE 12-4

Specialty-Flavored Carbonated Beverages Average Composition [a]

Type	Number of samples	Sugar, % °Brix	Carbonation, gas volume	Acid, %	pH
Birch beer	10	10.1	3.4	0.02	
Celery	2	9.0	3.2	0.05	
Cream	77	11.4	2.9	0.02	
Cream	15	11.2	2.6	0.02	
Ginger ale	16	8.0	3.1		3.0
golden	182	9.5	3.8	0.10	
golden	38	10.2	3.5	0.10	
pale dry	278	8.7	3.9	0.13	
pale dry	42	8.8	3.8	0.12	
Kola	64	10.7	3.1	0.08	
Kola	20	10.5	3.4	0.09	
Kola	30	10.1	2.8		2.6
Root beer	88	11.4	3.1	0.02	
Root beer	25	9.9	3.3	0.04	
Root beer	38	10.9	2.5		4.0
Sarsaparilla	27	9.7	3.5	0.03	

[a] After data of Toulouse, Hale, and Noling.

b. **Ginger-Ale Beverages**

A number of years ago, Toulouse [4] made a study of the composition of ginger ales. He pointed out that though it was customary to group together the two designations "pale" and "dry" as though they constituted a single definition, they were actually not to be considered as such. As mentioned on page 136, "pale" refers only to the color and "dry" is designed to convey a description of the amount of sweetness developed by both the sugar and acid content. Toulouse noted that it was unfortunate that these words were copied directly from their use in the designation of certain wines where "dry" means virtually sugar-free.

Toulouse examined several hundred samples of ginger ales and found the average composition of 160 samples labeled "golden" and 250 samples labeled "pale dry" to be that shown in Table 12-5.

The golden ginger ales appeared to be made with sugar concentrations between 8.5 and 9.5 per cent with a standard acid solution used at the rate of 1.25 ounces per gallon of 30°Bé, 1⅛ ounces per gallon of 27°Bé, and 1⅜ ounces per gallon of 32°Bé sirup.

Pale dry ginger ales appeared to be made with sugar contents of either 8 or 9 per cent with acid at the rate of 2¼ ounces per gallon of 30°Bé, 2 ounces per gallon of 27°Bé and 2¼ ounces per gallon of 32°Bé sirup.

Toulouse suggested the bottling of a "golden dry" ginger ale having the sugar content and color of a golden ginger ale with the acid concentration of a pale dry ginger ale.

It should be mentioned that a Belfast-type ginger ale is sweet and pungent.

TABLE 12-5

Composition of Ginger Ales [a]

Component	Golden (160 samples)	Pale dry (250 samples)
Sugar, %	9.490	8.550
Acid, grains/gal	59.100	73.800
Acid, %	0.010	0.013
Carbon dioxide, volumes	3.830	3.930

[a] After J. H. Toulouse, *Natl. Bottlers' Gaz.*, **52**, No. 618, 45 (1933).

3. FRUIT-FLAVORED BEVERAGES

Fruit-flavored beverages can be placed into two broad categories, from the point of view of composition, namely, those containing fruit juice with or without pulp and fruit cells, and other fruit-flavored beverages, including both artificially flavored ones and those flavored with natural fruit oils. Most members of the first group are citrus-flavored beverages, but other beverages in which fruit juice is commonly used are grape, apple, and pineapple.

Blangsted [5] cites typical formulations for carbonated citrus beverages. Thus a carbonated orange-flavored beverage, designed to contain 6 per cent orange juice, is prepared by making a beverage base and then a bottling sirup.

ORANGE-BEVERAGE BASE

Orange juice, concentrated, 65°Brix	1.0 gal
Citric acid solution, 50%	28.0 fl oz
Orange-oil emulsion, 15% cold-pressed (or equivalent in concentrated oil)	30.0 fl oz
FD&C Yellow No. 6 (sunset yellow FCF)	1.0 oz
Water	6.0 fl oz
Total	1.5 gal

Mix and homogenize to break up the pulp and fruit cells so that the particles will be small enough to pass through the bottling equipment without clogging any of the ports. The orange-beverage base is used to prepare a sirup as follows.

ORANGE-JUICE BOTTLERS' SIRUP

Orange beverage base	1 gal
Sugar	80 lb
Sodium benzoate solution (2 lb/gal)	21 fl oz
Water	8 gal, 7 pt

This formula will yield 16 gallons of orange-juice bottlers' sirup. A throw of 1.5 ounces is used for a 7-ounce bottle and, according to Blangsted, will yield a beverage having total solids, 13°Brix; total acidity, approximately 0.175 per cent, calculated as anhydrous citric acid by weight; orange-juice content, 6 per cent; volatile oil, 0.03 per cent; and sodium benzoate, 0.05 per cent.

Von Loesecke [6] cites an example of a carbonated lime soft drink.

LIME-JUICE BOTTLERS' SIRUP

Lime juice	30 fl oz
Sugar	4 lb 12 oz
Distilled lime oil (1% emulsion)	9.5 fl oz
Sodium benzoate	0.5 oz
Water, sufficient for	1 gal

This formula yields 1 gallon of lime-juice bottlers' sirup. A throw of 2 ounces is used per 12-ounce bottle. According to von Loesecke, the finished beverage contains total solids, 9°Brix; total acidity, approximately 0.4 per cent as citric acid; and lime-juice content, 5 per cent.

Blangsted [5] gives a typical example of a Tom Collins Mixer base:

TOM COLLINS MIXER BASE

Lemon juice, concentrated 6:1	0.75 gal
Lime juice, concentrated 6:1	0.50 gal
Lemon oil emulsion, 10% cold-pressed	6.00 fl oz
Lime oil emulsion, 10% cold-pressed	4.00 fl oz
Water	22.00 fl oz
Total	1.50 gal

After homogenization, the Tom Collins Mixer is made according to the following formulation:

TOM COLLINS MIXER BOTTLERS' SIRUP

Tom Collins Mixer base	1 gal
Sugar	54 lb
Sodium benzoate solution (2 lb/gal)	23 fl oz
Water	10 gal, 3 pt

This formulation will yield 15 gallons, 5 pints of Tom Collins Mixer bottlers' sirup. A throw of 6 fluid ounces is used for a 32-ounce bottle and will require a carbonation of 4 volumes of carbon dioxide. According to Blangsted, the resultant beverage will have a composition of total solids, 8°Brix; total acidity, approximately 0.35 per cent as anhydrous citric acid by weight; juice content, 6 per cent; volatile oil content, 0.0062 per cent; sodium benzoate, 0.05 per cent.

Still another example of such fruit-flavored carbonated beverages can be taken from von Loesecke,[6] namely, one for grapefruit flavor:

GRAPEFRUIT-JUICE BOTTLERS' SIRUP

Grapefruit juice, sweetened, 15°Brix	5.00 gal
Sugar	50.00 lb
Grapefruit oil emulsion (20% oil)	12.00 fl oz

Citric acid	23.00 oz
Naringin	0.75 oz
Sodium benzoate	3.25 oz
Water	3 gal, 2 pt

A throw of 3 ounces is used for a 12-ounce container and the recommended carbonation is 2 volumes of carbon dioxide. The finished beverage has a composition approximating: total acidity, 0.25 per cent as citric acid; volatile oil, 0.04 per cent; naringin (a natural bitter glucoside found in grapefruit), 100 ppm; and fruit juice, 10 per cent.

A number of years ago, Toulouse [7] made a study of the average composition of citrus-flavored beverages. The results of this study are shown in Table 12-6. The composition of six commercial, carbonated, orange-flavored soft drinks is given in Table 12-7.

The formulation of the fruit-flavored carbonated beverages other than citrus flavors was discussed in some detail in Chapters 7 and 8. In Table 12-8, the composition of fruit-flavored commercial carbonated beverages is detailed.

TABLE 12-6

*Citrus-Flavored Carbonated Beverages
Average Composition*

Type	Number of samples	Sugar, % °Brix	Carbonation gas volume	Acid, %	pH
Lime [a]	54	9.17	4.0	0.140	3.02
Lemon and lime [a]	20	11.04	3.2	0.175	3.01
Lemon and lime [b]	16	10.30	2.8		3.00
Lime [a]	54	11.10	3.7	0.228	2.90
Lemon [a]	348	11.18	3.2	0.120	3.07
Orange [a]	132	13.40	2.3	0.193	3.39
Orange [b]	38	12.90	1.7		3.10

[a] After J. H. Toulouse, *Ind. Eng. Chem.*, **26**, 767 (1934).

[b] A. W. Noling, *Soft Drinks in Cans* and *Supplement to Soft Drinks in Cans*, Indianapolis, Hurty-Peck & Co., 1954.

TABLE 12-7

Composition of Commercial Carbonated Orange-Flavored Beverages [a]

Sugar, % °Brix	Carbonation gas volume	Acid, % as citric	pH
14.3	1.2	0.174	2.65
12.8	1.3	0.215	3.20
13.9	1.0	0.165	3.50
14.0	2.1	0.171	2.60
13.0	1.9	0.123	3.80
12.0	1.9	0.128	3.10

[a] After Sara Blangsted, in D. K. Treesler and M. A. Joslyn, Eds., *The Chemistry and Technology of Fruit and Vegetable Juice Production*, New York, Avi Publishing, 1954, p. 608.

TABLE 12-8

Fruit-Flavored Carbonated Beverages Average Composition [a]

Type	Number of samples	Sugar, % °Brix	Carbonation gas volume	Acid, % as citric	pH
Apple	3	12.0	3.0	0.10	
Cherry	45	11.6	2.9	0.09	
Cherry	27	12.0	2.4		3.7
Grape	12	13.2	2.2	0.10	
Grape	89	13.5	2.1	0.10	
Grape	24	12.2	2.4		3.0
Peach	8	12.8	4.4	0.05	
Raspberry	8	11.2	2.3	0.10	
Strawberry	73	12.0	3.0	0.08	

[a] After data of Toulouse, Hale, and Noling.

TABLE 12-9

Composition of Commercial Fruit-Flavored Carbonated Beverages [a]

Type	Sugar, % °Brix	Carbonation gas volume	Acid, % as citric	pH
Blackberry	13.7	2.6	0.130	2.75
Blackberry	12.5	3.0	0.124	3.10
Blackberry	12.5	3.0	0.146	3.00
Blackberry	13.0	3.0	0.128	3.00
Grape	14.5	1.3	0.124	3.10
Lemonade	14.0	1.1	0.302	2.25
Lemon-lime	8.5	3.8	0.130	2.75
Lemon-lime	12.6	2.4	0.097	2.95
Lemon-lime	13.2	3.0	0.347	3.07
Lime	13.3	2.6	0.208	2.70
Raspberry	11.0	3.0	0.108	3.40
Raspberry	12.3	3.0	0.134	3.00
Tom Collins	7.4	4.0	0.365	3.20

[a] After Sara Blangsted, in D. K. Tressler and M. A. Joslyn, Eds., *The Chemistry and Technology of Fruit and Vegetable Juice Production*. New York, Avi Publishing, 1954, p. 609.

LITERATURE CITED

1. J. H. Toulouse, *Mineral Water Trade Rev.*, **1933**, 276.
2. A. W. Noling, Ed., *Soft Drinks in Cans* and *Supplement to Soft Drinks in Cans*, Indianapolis, Hurty-Peck & Co., 1954.
3. G. B. Beattie, *Perfumery Essent. Oil Rec.*, **47**, No. 12, 437 (1956).
4. J. H. Toulouse, *Natl. Bottlers' Gaz.*, **52**, No. 618, 45 (1933).
5. S. Blangsted, in D. K. Tressler and M. A. Joslyn, Eds., *Chemistry and Technology of Fruit and Vegetable Juice Production*, New York, Avi Publishing, 1954.
6. H. W. von Loesecke, *Outlines of Food Technology*, 2nd Ed., New York, Reinhold, 1949.
7. J. H. Toulouse, *Ind. Eng. Chem.*, **26**, 767 (1934).

SELECTED BIBLIOGRAPHY

J. F. Hale, in M. B. Jacobs, Ed., *Chemistry and Technology of Food and Food Products*, 1st Ed., Vol. II, Chapter XXIII, New York, Interscience Publishers, 1944.

H. W. von Loesecke, *Outlines of Food Technology*, 2nd Ed., New York, Reinhold, 1949.

A. W. Noling, Ed., *Soft Drinks in Cans and Supplement to Soft Drinks in Cans*, Indianapolis, Hurty-Peck & Co., 1954.

R. Kühles, *Handbuch der Mineralwasser-Industrie*, Lübeck, Antäus-Verlag, 1947.

S. Blangsted, in D. K. Tressler and M. A. Joslyn, Eds., *Chemistry and Technology of Fruit and Vegetable Juice Production*, New York, Avi Publishing, 1954.

Chapter 13

PLANT LAYOUT AND SANITATION

Since many factors of adequate plant layout are closely tied in with good plant-sanitation practice, these two aspects may well be considered together. Time and again at various points in the preceding chapters, the necessity for keeping the beverage plant in a sanitary condition was stressed. This cannot be emphasized too strongly. The proper plant layout and proper plant housekeeping are important for good sanitation.

1. PLANT LAYOUT

It is essential that sound engineering principles are followed in the plant layout and in the routing schedule. Unless this is arranged properly, much unnecessary work will have to be done, resulting in an increase in production costs. One could design a plant theoretically that would perhaps solve many, if not all, such engineering difficulties. This is very difficult to do in actual practice, because many factors enter into the choice and arrangement of a plant. Among these are proximity to sources of shipment of raw materials, for instance, location near a railroad; proximity to major distribution points, that is, the consuming area served; zoning, that is the permissible operation and the character of the neighborhood; water supply; availability of labor; etc. Sometimes just one of these factors may loom so large that the plant must be placed and arranged so that the conditions of that one factor are met. Indeed, the chance of purchasing the required land and subsequently designing and building a plant or renting a plant may be out of the question for all of this may have to be subordinated to the controlling factor.

The subject of plant layout has been discussed thoroughly in many carbonated-beverage and food journal articles. It would be best here to call attention to certain major engineering aspects. Some of these may have been mentioned in other chapters, but will bear repetition here.

If the sirup room is located on a balcony or an upper floor of a beverage plant and granulated sugar is being used, it is more convenient, as was pointed out in Chapter 2, to locate the sugar mixing tank on the ground floor or the floor where the sugar is delivered and to pump the sirup to the sirup room than to carry or transport bags of sugar to the sirup room. This is true even if bulk sugar is received by carload or truck, for then sirup-making equipment can be attached, as described in Chapter 2, and the resultant sirup can be pumped to the sirup storage.

If the sirup room is located on a balcony or an upper story, then advantage can be taken of gravity flow of sirup to the sirup cooler and from there to the filler. In a one-story arrangement, the finished sirup has to be pumped directly to the filler, or to the premix carbonator, or to a supply tank. In the case of liquid-sugar delivery, the liquid sugar must be pumped to a supply tank.

When sirup is not manufactured on the premises, but is received as a finished flavored and acidulated sirup in drums from a major manufacturer, it is still advantageous to have the drums placed on a balcony or second or upper floor. The position of the drums and for that matter the sirup tanks should be so adjusted that a sirup cooler can be used and a minimum amount of piping is required to take the sirup to the filler or the premix carbonator.

In a two-story system, the sirup cooler can be located as a floor-ceiling arrangement (Figure 13) and advantage can be taken, as mentioned, of gravity flow. If a one-floor operation is being followed (Figure 14), then a slight pressure of carbon dioxide must be maintained over the sirup cooler to force the chilled sirup into the filler.

If a system using proportionating equipment, like a Syncrometer, is being used or contemplated, then a single-story operation (Figure 15) can be followed, since such a system is equipped with pumps to feed both water and sirup to the filler.

PLANT LAYOUT AND SANITATION

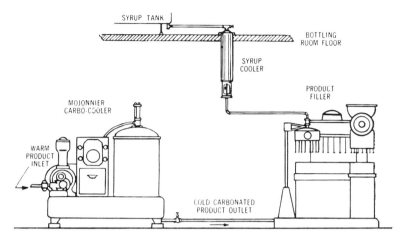

FIGURE 13. Conventional Filling, Using Carbonated Water at 34°F and Sirup at 50°F; Two-Floor Operation (Courtesy of the Mojonnier Bros. Co., Chicago, Ill.)

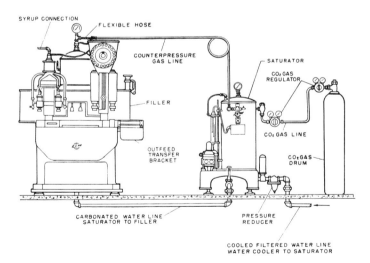

FIGURE 14. Diagram of Water and Gas Connections to the Machines in Conventional Filling (Courtesy of the Crown Cork & Seal Co., Inc., Philadelphia, Pa.)

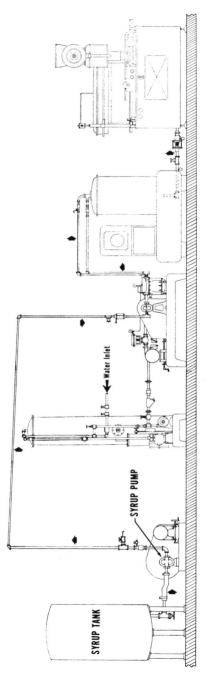

FIGURE 15. One-Floor Operation of a Syn-Cro-Mix Installation (Courtesy of the Geo. J. Meyer Manufacturing Co., Milwaukee, Wis.)

PLANT LAYOUT AND SANITATION

The deaerator can be placed in a basement or in other sections of the plant where most convenient.

The water-treatment equipment, ultraviolet sterilizer for water, if used, water cooler, sirup cooler, water deaerator, water carbonator, or premix carbonator, filler and capper, or filler and can sealer should be arranged so that there are no long runs of pipe. This avoids the warming up of the carbonated water or the sirup in the run to the filler which may result in foaming at the filler. Shorter lines also mean less cleaning time. However, the lines must not be so short that there is difficulty in getting to or around the equipment for cleaning or repair.

It is relatively common to place all of such equipment on the ground or first floor and to use large glass windows so that all the operations mentioned are visible to the public. When this is done, it is also advisable to have an air-conditioning system. Such a system will not only assist in maintaining cleanliness, but by use of properly humidified air avoids condensation on the windows.

For loading and unloading, if a basement is available, a satisfactory arrangement is one in which the empty bottles and their cases are taken from a truck at the first or ground floor level, placed on a conveyor which carries them to a case inverter, the bottles being automatically unloaded and passed on to a washer. They then go to the filler on the first floor and, after filling, capping, and labeling, are returned to the basement to meet the cleaned cases. After packing, the filled cases are conveyed up to the first or ground floor to be placed on trucks.

The truck loading and unloading facilities must be carefully considered and arranged. A principal factor in the arrangement is ease in handling cases to and from conveyor lines. The conveying lines should run readily from the returning dock to the used-bottle storage and from there to the washer feeds. The cases should have free run to the case cleaner.

Sometimes a location can be obtained where the bottling operations are on a street level which is one story higher than the rear of the building which is on a lower street level. Under such circumstances the trucks can load and unload on the same level as the bottle storage and bottle sorting.

In a large plant, palletizing handling should be considered. In this case, adequate space must be allowed for the movement of the lift trucks that move the pallets. Some thought must also be given to adequate ventilation when gasoline-powered lift trucks are used. Pallet-bodied trucks must be employed.

There are four major advantages of using pallets for loading and unloading. (1) There is a marked reduction in labor costs, for far less labor is required than in manual handling. (2) Since loading and unloading with pallets is much faster, trucks can be loaded and unloaded faster, thus cutting the waiting and turn-around time of the trucks, indeed such palletizing may actually reduce the number of trucks needed for serving consumers. (3) Maximum use of storage space can be obtained for the stack heights of cases can be increased considerably over that obtainable with manual handling. (4) The actual handling operations can be materially reduced. Thus in any beverage handling situation, the manual operation comprises the following steps: The empty bottles and their cases must be unloaded; the empties must be fed to conveyors for stacking or must be stacked directly; the empties must be fed to conveyors to the soakers; the full goods must be stacked; and finally the full goods must be loaded. With palletizing most of these operations can be reduced merely to placing empty cases on conveyors to the soakers and so on to the fillers and removal of the filled cases for placement on pallets to be subsequently loaded on trucks.

One additional advantage of using pallets is a reduction in bottle breakage as compared with such losses sustained when the bottles are handled manually.

2. PLANT HOUSEKEEPING

Plant housekeeping may be defined as those administrative factors and details which result in satisfactory plant sanitation and cleanliness. The successful operation of any carbonated-beverage plant depends in full measure on good plant-housekeeping practices, that is, on a complete sanitation and cleaning program. It would be best if such a program were placed by management in the hands of a sanitarian or preferably a food technologist, but it must be realized that many beverage plants cannot afford such a full-time employee. It is then incumbent on the manager or operator of such a plant to make certain that proper sanitation is maintained and to impress all employees with the need for this.

The first requirement in this respect is that the building housing the plant is built or remodeled with the thought in mind that it is a food establishment and that all of the sanitary requirements of such establishments must be met. Thus all of the floors should be made of dense durable material with proper drainage facilities so that they can be washed. The walls should be made of or covered with a material that can be washed also. All the corners and wall and floor junctions should be coved so that they can be easily washed and infestation is not invited by cracks or difficult corners.

Buchanan [1] has summarized many of these factors for the sirup room. He recommends that this room should have (1) plenty of light, (2) tile floors sloped for good drainage, (3) tile walls to a height of at least 6 feet from the floor, (4) smooth painted ceiling and walls above the tiling, (5) well screened windows and doors; and (6) air conditioning.

The basic purpose of the suggestions made in the preceding paragraphs is making cleaning easy.

One of the most important aspects of proper plant housekeeping, and this has been mentioned in the discussion concerning adequate plant layout, is the proper placement of all equipment and machines. This cannot be stressed too strongly. Abrahamson [2] has pointed out that all machines and equipment must be placed and installed in such a manner that no part of the equipment is inaccessible for cleaning. The equipment should be placed and installed so that there is free and ready access around it for inspection, cleaning, and repair.

All of the equipment should be of sanitary construction. In the United States, there has been considerable progress in the manufacture of equipment having sanitary design. Possibly the milk and dairy industry lead in this respect, but the carbonated beverage industry has also made contributions.

Despite sanitary construction, all machines must be cleaned, that is, sanitary construction is no substitute for cleanliness, but makes cleaning easier and faster.

In general, the equipment should be made of stainless steel. Stainless steel piping with sanitary fittings, which facilitate dismantling and reassembly, also contribute to the ease of cleansing and are probably best for the lines leading into and from the sirup room. Pyrex-glass piping has been used successfully for lines to and from sirup tanks. Plastic tubing and hose have been used for lines to filters.

3. BOTTLE WASHING

It has been pointed out by a number of authorities in the carbonated-beverage industry that bottle washing is one of the most important steps in beverage production. Generally bottle washing comprises soaking or flushing the bottles with a solution of caustic soda (sodium hydroxide) in combination with other cleansing agents, such as soda ash (sodium carbonate), sodium aluminate, trisodium phosphate, various polyphosphates, sodium metasilicate, sodium borate, and other alkaline compounds, followed by a thorough scrubbing of the bottles both inside and out, and finally rinsing with potable water equivalent to that used in making the beverage, but preferably softened.

a. Criteria

Hale [3] stressed that the criteria for acceptable bottle washing are that the bottles be (a) made sterile, (b) rinsed free of the chemical or sterilizing agents used, (c) of acceptable appearance, and (d) of good mechanical strength. In order to achieve these criteria, the factors to be controlled in the bottle-washing operation are (1) adequate concentration of sterilizing and detergent agents, (2) proper composition of the agents, (3) adequate temperature of the washing solution, (4) sufficient length of time of exposure of the bottles to the washing and sterilizing agents, (5) proper rinsing water, and (6) proper maintenance of the bottle-washing equipment.

The American Bottlers of Carbonated Beverages has made an intensive study of these factors and of bottle-washing compounds. These studies have shown that the most widely recognized minimum conditions for complete and thorough cleansing and sterilizing of bottles to be used for carbonated beverages are (a) the solutions should contain 3 per cent of total alkali, of which not less than 60 per cent is caustic soda (sodium hydroxide), that is 1.8 per cent of the total alkali should be sodium hydroxide; (b) the bottles should be soaked for a minimum of 5 minutes in the sterilizing and cleansing solution; and (c) the temperature of the washing solution must be kept at 130°F as a minimum. These conditions will control the first four factors mentioned.

It must be pointed out that the portion of the alkalinity that is due to the sodium hydroxide is most important. Thus merely having 3 per cent total alkalinity is not sufficient to produce sterility. There must also be at least 1.8 per cent of free sodium hydroxide. The methods for testing for total alkalinity and free sodium hydroxide are detailed in Chapter 15. It is best to use these methods. An approximate check on the amount of caustic can be made by determining the specific gravity in degrees Baumé. Years ago Buchanan, Levine, and McKelvey [4] recommended a 3.0 to 3.5 per cent alkali solution and discarding the solution when it became dark or black.

b. **Bottle-Washing Machines**

Bottle washing machines are of a number of types. They may be put into three categories, namely immersion or soaker washers, immersion or soaker washers in combination with brushing, and so-called Hydro washers in which the bottles are subjected to a combination of scouring and washing action by means of powerful jets. To meet the criteria mentioned, bottle-washing machines should have a soaking operation, a brushing step, and a final rinsing. Equipment is available with a prerinse section and both inside and outside brushing with nonmetallic brushes after use of special detergents. Bottle-washing machines are of different sizes, the principal sizes being machines capable of washing ½ pint, pint, and quart bottles, respectively.

One of the more common bottle-washing machines is the four-compartment, double-end machine of Figure 16. The bottles are placed on an accumulative loader and are automatically fed by a device which handles the bottles separately so that they do not scuff and crowd. Some machines are equipped with unscrambler loaders and others with case unloaders. The bottles are separated by stops and are loaded individually into bottle carriers in a horizontal position.

The second step is a prerinse in which the bottles are given a thorough flushing. In some machines, this prerinse is not given until the bottles are in an almost vertical position. This permits emptying of liquid and some debris before the prerinsing. In other machines, the bottles are prerinsed in a horizontal position immediately after being loaded into the bottle carrier. This gives the bottles additional opportunity to drain completely into a catch basin before they reach the first caustic compartment and thus prevents the dilution of the caustic by the prerinse water.

The bottles pass into the first compartment containing usually

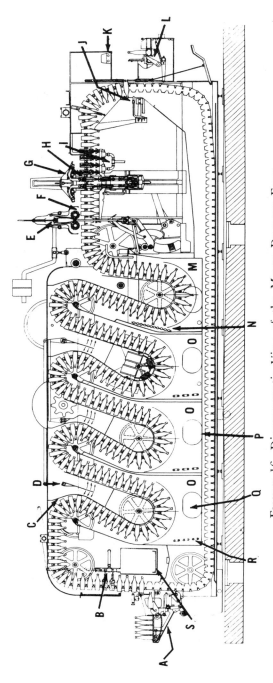

FIGURE 16. Diagrammatic View of the Meyer Dumore Four-Compartment Bottle Cleaner (Courtesy of the Geo. J. Meyer Manufacturing Co., Milwaukee, Wis.)

A—Loading mechanism
B—Horizontal prerinse
C—Bottle guards
D—Agitator
E—Bottom brush
F—Outside brushes
G—Top clamp safety
H—Two-rows of inside brushes
I—Final rinse
J—Push-out finger
K—Inspection light
L—Automatic unloading
M—Myrex or fresh-water compartment
N—Hot-water heater
O—Caustic compartments
P—Large manholes flush with bottom of tank
Q—Large space for labels
R—Steam heater
S—Drain pan

3 per cent caustic solution, or 3 per cent caustic solution plus a bottle-cleaning detergent, such as sodium aluminate or other cleansing components, at a temperature of 110°F and from there to the second compartment which contains either a 3.0 or 2.5 per cent caustic solution, or an equivalent washing mixture at 150°F. The bottles must, in order to meet the criteria mentioned, be immersed in these solutions, that is, have a total residence time in these solutions, of at least 5 minutes. In the third compartment, the caustic concentration is less and the temperature is reduced to 130°F. Each of the first three compartments has a relatively large space at the bottom for the accumulation of label and other debris and each compartment is separated from the others by a double air space to insulate it, thus reducing heat transfer. In the first three compartments, the liquid level is at about the axle of the sheave wheels.

The fourth tank or compartment contains fresh water, or water with chelating, emulsifying, and detergent agents, at a temperature of 100°F to assist in removing traces of grease and oil not removed by the caustic treatment. One of these commercial cleansing mixtures is "Myrex."

The bottles pass from the fourth compartment, with some of the special cleansing agent adhering to the surfaces, if such a mixture is used, to an outside brushing section at which point each bottle is raised and brushed, the brushes rising and falling with the bottle thus increasing the brushing time, and then to an inside brushing section where they are brushed at high speed preferably with non-metallic brushes. If special cleansing agents are used, a foamy lather is formed which assists in the removal of all dirt, grease, and oil that may still adhere to the glass. Such cleansing agents assist the brushes in cleansing themselves so that they do not contaminate the following bottles. In some machines, if a brush hits an obstruction, the machine is stopped. Finally the bottles are rinsed by a series of jet rinsers that deliver a spray of water. The bottles are then pushed out by knockout fingers with a gentle motion when they reach the discharge port and are automatically placed in an upright position so that they can be inspected as they leave the soaker.

c. Washing Solutions

In subsection 13.3.a of this chapter, the principal substances used in making soaker solutions, the proper caustic concentrations, temperature of the washing solutions, and time of soaking were stated.

A large-scale manufacturer of bottle washing machines recommends the following bottle-washing chemicals in the order of preference: (1) a mixture of caustic soda, containing straight 76 per cent caustic soda, and sodium aluminate, specifically Nalco 680, (2) straight 76 per cent caustic soda, and (3) 76 per cent caustic soda with a small amount of soda ash, that is, anhydrous sodium carbonate.

In the experience of this firm, the controlled use of sodium aluminate in combination with caustic soda will yield clean, sterile bottles and will at the same time prevent scuffing and marring of the bottles being washed. The Geo. J. Meyer Manufacturing Co.,[5] recommends the elimination of all phosphates and silicates from the alkali compartments of bottle washers.

(1) Charging Alkali Compartments

The charging of the alkali compartments of the soaker is most important. In order to do this properly, the capacity of each compartment must be known in either cubic feet or in gallons. By use of Table 13-1, the amount of 76 per cent caustic soda required for a compartment of a given capacity can be calculated. It is important to dissolve the caustic in a separate vessel to avoid formation of a glassy deposit on the bottom of the tank (See section 13.3.d.).

TABLE 13-1

Caustic Soda (76%) Soaker Solutions

Sodium hydroxide % desired	Caustic soda required	
	pound/gallon	pounds/cubic feet
0.5	0.05	0.32
1.0	0.09	0.63
1.5	0.13	0.95
2.0	0.17	1.27
2.5	0.21	1.60
3.0	0.26	1.93
3.5	0.30	2.27
4.0	0.35	2.60
4.5	0.39	2.94
5.0	0.44	3.29

By the use of Table 13-2 the amount of sodium aluminate to be added to caustic soda solutions, if desired, can be calculated for a given compartment capacity and a given percentage of sodium aluminate. It must be remembered that, if bottles having aluminum foil labels are being washed, the caustic soda will dissolve the aluminum and there will be no need to add sodium aluminate. Indeed, it will be necessary to keep a close check on the caustic concentration in order to maintain it at the proper strength. This is discussed in detail in section 13.3.e.

TABLE 13-2

Sodium Aluminate[a] Requirements for Soaker Solutions[b]

Sodium aluminate	Sodium aluminate required	
% desired	pound/gallon	pound/cubic foot
0.1	0.008	0.06
0.2	0.017	0.12
0.3	0.025	0.18
0.4	0.034	0.25
0.5	0.043	0.32

[a] Nalco 680.
[b] After W. C. Cooper, "Efficient Bottle Washing," Bull. No. B 139, Cudahy, Wis., Geo. J. Meyer Manufacturing Co., 1957.

In Table 13-3, the caustic concentration and temperature suggested for the different compartments of three-, four-, five-, and six-compartment soaker washers are tabulated and in Table 13-4, the caustic concentration, sodium aluminate concentration, and temperature suggested for the corresponding compartments of soaker washers are also listed.

TABLE 13-3

Caustic Concentrations and Temperatures of Soaker Compartments[a]

Compartment											
1		2		3		4		5		6	
NaOH %	°F	NaOH %	°F	NaOH %	°F	NaOH %	°F	NaOH %	°F	NaOH %	°F
3	110	2.5	150	None[b]	100						
3	110	2.5	150	1.5	130	None	100				
3	110	2.5	150	2.5	170	1.5	135	None	100		
3	110	2.5	150	2.5	170	1.5	170	1.0	135	None	100

[a] After W. C. Cooper, "Efficient Bottle Washing," Bull. No. B 139, Cudahy, Wis., Geo. J. Meyer Manufacturing Co., 1957.
[b] None indicates that fresh water is used.

With the progress of washing, the washing solutions gradually lose caustic strength. This is due to the reaction of the sodium hydroxide with labels, dirt, and carry-over of caustic adhering to the bottles. It is, therefore, necessary to check the causticity from time to time and add additional caustic as required. In order to do this properly, the washing solution should be analyzed. Simple methods of analysis are detailed in Chapter 15.7.

TABLE 13-4

Caustic Concentrations, Aluminate Concentrations, and Temperatures of Soaker Washer Compartments [a]

Compartment No.					
1	2	3	4	5	6
110°F	130°F	100°F			
Caustic 3%	Caustic 2½%	Fresh			
No. 680 [b] 5/10%	No. 680 2/10%	Water			
110°F	150°F	130°F	100°F		
Caustic 3%	Caustic 2½%	Caustic 1½%	Fresh		
No. 680 5/10%	No. 680 3/10%	No. 680 2/10%	Water		
110°F	150°F	170°F	135°F	100°F	
Caustic 3%	Caustic 2½%	Caustic 2½%	Caustic 1½%	Fresh	
No. 680 5/10%	No. 680 3/10%	No. 680 2/10%	No. 680 2/10%	Water	
110°F	150°F	170°F	170°F	135°F	100°F
Caustic 3%	Caustic 2½%	Caustic 2½%	Caustic 1½%	Caustic 1½%	Fresh
No. 680 5/10%	No. 680 3/10%	No. 680 2/10%	No. 680 2/10%	No. 680 1/10%	Water

[a] W. C. Cooper, "Efficient Bottle Washing," Bull. No. B 139, Cudahy, Wis., Geo. J. Meyer Manufacturing Co., 1957.
[b] Nalco 680 (sodium aluminate).

Still another and possibly simpler method of computing the amount of caustic soda required is to use Table 13-5, in which the pounds of caustic required are related to the tank capacity.

It is used as follows: Assume that the capacity of the tank is 573 gallons and it is to be charged with a fresh solution of caustic soda containing 3 per cent of sodium hydroxide. Look under the column marked 3 per cent and follow it down to the figure aligned with 500 gallons, record the figure found. Repeat this with 70 and 3 gallons, respectively, under the 3 per cent column and also record the figures. This will then appear as:

	gallons	pounds
	500	130.00
	70	18.20
	3	0.78
Total	573	148.98

Thus 148.98 pounds of caustic soda are required to charge a compartment having a capacity of 573 gallons with 3 per cent caustic soda.

(2) Strengthening Washing Solutions

It has been stressed that the caustic strength of the washing solutions must be maintained at required levels in order to have efficient washing. The following procedure may be used to bring caustic solutions up to proper strength. If sodium aluminate is being used or bottles with aluminum labels are being washed, refer to the section on label removal for the method of strengthening such caustic solutions.

Add sufficient water to fill each compartment to within about 3 inches below the overflow level when there are no bottles in the soaker and the machine is stopped. Operate the soaker for a few minutes in order to mix the contents thoroughly and then draw a sample from each compartment as detailed in Chapter 15.7. Make an analysis of each sample. By reference to Tables 13-1, 2, 3, and 4, the amount of caustic soda or sodium aluminate required can be calculated. Thus subtracting the value found by analysis from the standard value for that compartment, Tables 13-3 and 4 give the amount in percentage required to bring the solution to proper strength and Tables 13-1 and 2 give the amount in weight.

Cooper [5] gives the following example. Suppose a compartment of 100 cubic feet capacity is required to contain 2.5 per cent of sodium hydroxide (caustic) and 0.3 per cent of sodium aluminate (Nalco 680). The analysis either by titration with acid or with tablets shows that the washing solution actually contains 2.0 per cent of sodium hydroxide and 0.2 per cent of sodium aluminate, then:

	$NaOH$, %	$Na\ Aluminate$, %
Required percentage	2.5	0.3
Percentage found by analysis	2.0	0.2
Needed percentage	0.5	0.1

By reference to Table 13-1, it can be seen that 0.5 per cent of caustic is equivalent to 0.32 pound of caustic per cubic foot, thus 100 cubic feet (the capacity of the compartment) will need 0.32 × 100 or 32 pounds of caustic soda to bring the solution up to strength.

TABLE 13-5
Caustic Control Chart [a]

Capacity of tank gallons	Per cent of caustic wanted													
	0.1	0.2	0.3	0.4	0.5	0.6	0.7	0.8	0.9	1.0	2.0	3.0	4.0	5.0
	Pounds of caustic needed will be													
1000	9.0	18.0	27.0	36.0	45.0	54.0	63.0	72.0	81.0	90.0	170.0	260.0	350.0	440.0
900	8.1	16.2	24.3	32.4	40.5	48.6	56.7	64.8	72.9	81.0	153.0	234.0	315.0	396.0
800	7.2	14.4	21.6	28.8	36.0	43.2	50.4	57.6	64.8	72.0	136.0	208.0	280.0	352.0
700	6.3	12.6	18.9	25.0	31.5	37.8	44.1	50.4	56.7	63.0	119.0	182.0	245.0	308.0
600	5.4	10.8	16.2	21.6	27.0	32.4	37.8	43.2	48.6	54.0	102.0	156.0	210.0	264.0
500	4.5	9.0	13.5	18.0	22.5	27.0	31.5	36.0	40.5	45.0	85.0	130.0	175.0	220.0
400	3.6	7.2	10.8	14.4	18.0	21.6	25.2	28.8	32.4	36.0	68.0	104.0	140.0	176.0
300	2.7	5.4	8.1	10.8	13.5	16.2	18.9	21.6	24.3	27.0	51.0	78.0	105.0	132.0
200	1.8	3.6	5.4	7.2	9.0	10.8	12.6	14.4	16.2	18.0	34.0	52.0	70.0	88.0
100	0.9	1.8	2.7	3.6	4.5	5.4	6.3	7.2	8.1	9.0	17.0	26.0	35.0	44.0
90	0.8	1.6	2.4	3.2	4.1	4.9	5.7	6.5	7.3	8.1	15.3	23.4	31.5	39.6
80	0.7	1.4	2.2	2.9	3.6	4.3	5.0	5.8	6.5	7.2	13.6	20.8	28.0	35.2
70	0.6	1.3	1.9	2.5	3.2	3.8	4.4	5.0	5.7	6.3	11.9	18.2	24.5	30.8
60	0.5	1.1	1.6	2.2	2.7	3.2	3.8	4.3	4.9	5.4	10.2	15.6	21.0	26.4
50	0.5	0.9	1.4	1.8	2.3	2.7	3.2	3.6	4.1	4.5	8.5	13.0	17.5	22.0
40	0.4	0.7	1.1	1.4	1.8	2.2	2.5	2.9	3.2	3.6	6.8	10.4	14.0	17.6
30	0.3	0.5	0.8	1.1	1.4	1.6	1.9	2.2	2.4	2.7	5.1	7.8	10.5	13.2
20	0.2	0.4	0.5	0.7	0.9	1.1	1.3	1.4	1.6	1.8	3.4	5.2	7.0	8.8
10	0.1	0.2	0.3	0.4	0.5	0.5	0.6	0.7	0.8	0.9	1.7	2.6	3.5	4.4
9	0.08	0.16	0.24	0.32	0.41	0.49	0.57	0.65	0.73	0.81	1.53	2.34	3.15	3.96
8	0.07	0.14	0.22	0.29	0.36	0.43	0.50	0.58	0.65	0.72	1.36	2.08	2.80	3.52
7	0.06	0.13	0.19	0.25	0.32	0.38	0.44	0.50	0.57	0.63	1.19	1.82	2.45	3.08
6	0.05	0.11	0.16	0.22	0.27	0.32	0.38	0.43	0.49	0.54	1.02	1.56	2.10	2.64
5	0.05	0.09	0.14	0.18	0.23	0.27	0.32	0.36	0.41	0.45	0.85	1.30	1.75	2.20
4	0.04	0.07	0.11	0.14	0.18	0.22	0.25	0.29	0.32	0.36	0.68	1.04	1.40	1.76
3	0.03	0.05	0.08	0.11	0.14	0.16	0.19	0.22	0.24	0.27	0.51	0.78	1.05	1.32
2	0.02	0.04	0.05	0.07	0.09	0.11	0.13	0.14	0.16	0.18	0.34	0.52	0.70	0.88
1	0.01	0.02	0.03	0.04	0.05	0.05	0.06	0.07	0.08	0.09	0.17	0.26	0.35	0.44

[a] Courtesy Geo. J. Meyer Manufacturing Co., Cudahy, Wis.

By reference to Table 13-2, it can be seen that 0.1 per cent of Nalco 680 is equivalent to 0.06 pound per cubic foot, thus 100 cubic feet, that is, the capacity of the compartment would require 0.06 × 100 or 6 pounds of the sodium aluminate to bring the washing solution up to proper strength for this component.

An analogous method of computation can be used if the capacity of the compartments is known in terms of gallons instead of cubic feet.

Table 13-5 can also be used to calculate the amount of caustic that needs to be added to bring a caustic solution up to proper strength. Check the causticity of the compartment in question in the manner detailed in Chapter 15. Assume that it is found to contain 2.7 per cent of caustic and the standard required for that compartment is 3 per cent. Then subtracting 2.7 per cent from 3.0 per cent indicates that 0.3 per cent additional caustic soda is required. Assume again that a tank of a capacity of 573 gallons is to be strengthened. Look under the column headed 0.3 per cent and follow it down to the figure aligned with 500 gallons. Record the figure found. Repeat this with 70 and 3 gallons, respectively, under the 0.3 per cent column and record these figures, too, as follows:

	gallons	pounds
	500	13.50
	70	1.90
	3	0.08
Total	573	15.48

Thus 15.48 pounds of caustic soda are required to bring the compartment in question, having a capacity of 573 gallons, up to a concentration of 3 per cent caustic soda.

d. Predissolving Caustic Soda

The concentration of sodium hydroxide in the soaker solution is a most important factor in the washing of the beverage bottles and in producing sanitary bottles. Korab[6] of American Bottlers of Carbonated Beverages has pointed out that the conventional method of adding caustic soda to a solution, which consists of pouring of a predetermined number of pounds of dry caustic with the aid of buckets or a shovel into the alkali tank of the soaker, is not only time-consuming, but is also a dangerous method of handling the caustic soda. Considering the time required, it is a costly labor practice. Because of the fact that time is required for dry caustic to dissolve completely in alkali solution or in water and to be evenly distributed throughout the resultant solution, a test cannot be made immediately to determine the strength of the final solution. This is an additional cause for delay and an added disadvantage. Almost invariably some of the caustic soda will sink to the bottom of the tank and form a solid mass which may take hours to dissolve. If only the minimum alkaline strength is being maintained for sterilization, the caustic concentration will not reach its full value as long as there is any undissolved sodium hydroxide.

The practice of predissolving caustic soda before addition to the soaker not only saves labor but also saves material for no excess sodium hydroxide will have to be used, since the concentrated caustic soda solution will disperse rapidly throughout the washer solution while the soaker is in operation. The caustic soda solutions can be kept more closely to the most efficient operating strength because of the ease of adding the make-up solutions during the operating day. This method of adding caustic is also far less hazardous to the operator.

Korab illustrates the man-hours saved by predissolving with the following example. The amount of caustic used per year by a beverage manufacturer was approximately 51,000 pounds. The man-hours per week required to clean, lubricate, and add dry caustic soda to two soakers was 40. After changing over to the predissolving system of adding caustic soda solution, only 15 man-hours were required. The 25 man-hours saved per week was due directly to the ease of preparation of the concentrated caustic soda solution and the use of pumping to transfer it to the washer.

A number of predissolving systems can be arranged. One of the simplest is shown in Figure 17. The mixing tank A should have a larger capacity than the amount of caustic solution desired at any given period. For instance, we can take a tank which has a total capacity of not less than 100 gallons. Any shape can be used, but it should preferably be located as close to the soaker as practicable. As a preliminary step, the mixing tank should be filled with water and the gage glass B should be graduated to read in 2.5- and 5-gallon divisions up to 90 gallons. To prepare the con-

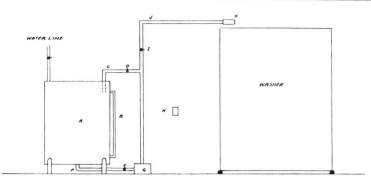

FIGURE 17. Simple Caustic Soda Predissolving System (Courtesy of the American Bottlers of Carbonated Beverages)

A—Mixing, measuring, and storage tank
B—Gage glass, marked to indicate gallons
C—Flow return pipe
D—Iron gate valve flow return
E—Iron gate valve
F—One-inch black-iron feed line to pump
G—Pump; iron-construction, open-type impeller
H—Motor-control switch; locate close to mixing tank
I—Iron gate valve
J—One-inch black-iron feed line to washer
K—Hose of proper length to reach all alkali compartments

centrated caustic solution, 70 gallons of water is first put into the mixing tank. With valve I closed and valves D and E opened, the pump is turned on, causing the water to flow through pipe F up to pipe C and back into the mixing tank. Slowly, 400 pounds of caustic soda is added. This will require about 10 to 15 minutes. The pump is permitted to circulate the mixture until all of the caustic is dissolved, which will take 20 to 30 minutes.

Immediately after the caustic is dissolved, the alkali solution will have a total volume of 85 gallons, as a result of the additional volume attributable to the caustic soda and to the expansion in volume resulting from the increase in temperature due to the heat of solution of the caustic in water. After cooling to 68°F, the total volume will decrease to 82 gallons. The concentrated caustic solution has a strength of 40.7 per cent and contains 4.88 pounds of caustic soda per gallon of solution. Since caustic hydrate begins to precipitate from a 40.7 per cent solution at 59°F, the stock solution should be stored at temperatures above 60°F.

The percentage of sodium hydroxide present in the concentrated

TABLE 13-6

Scheme for Preparing Soaker Solutions

(using predissolved caustic soda system [a])

Soaker tank capacity in gallons [b]	Percentage Caustic Content of Dry Compound												
	100 [c]			90 [c]				80 [c]					
	Gallons of concentrated caustic solution to add to give desired caustic strength in soaker [d]												
	1.5%	2.0%	2.5%	3.0%	1.5%	2.0%	2.5%	3.0%	1.5%	2.0%	2.5%	3.0%	
100	2.5	3.3	4.1	5.0	2.7	3.7	4.6	5.5	3.0	4.0	5.0	6.0	
125	3.1	4.2	5.2	6.2	3.4	4.8	5.7	6.8	3.7	5.0	6.2	7.5	
175	4.4	5.8	7.2	8.7	4.8	6.4	8.0	9.6	5.2	7.0	8.7	10.5	
200	5.0	6.6	8.3	10.0	5.5	7.3	9.1	10.9	8.0	8.0	9.9	11.9	
225	5.8	7.4	9.3	11.2	6.6	8.2	10.3	12.3	7.0	9.0	11.2	13.4	
275	6.8	9.1	11.4	13.7	7.5	10.0	12.5	15.0	8.3	10.9	13.7	16.4	
300	7.4	10.0	12.4	15.0	8.2	11.0	13.7	16.5	8.9	12.0	14.9	17.7	
325	8.1	10.8	13.5	16.2	8.9	11.9	14.8	17.8	9.8	13.0	16.1	19.5	
375	9.3	12.4	15.5	18.6	10.2	13.6	17.0	20.4	11.2	14.9	18.6	22.6	
400	9.5	13.3	16.6	19.9	10.9	14.6	18.3	21.8	13.0	15.9	20.0	24.0	
425	10.5	14.1	17.6	21.1	11.6	15.5	19.3	23.3	12.7	16.9	21.1	25.4	
475	11.8	15.8	19.7	23.6	13.0	17.5	21.7	26.0	14.2	18.9	23.6	28.6	
500	12.4	16.6	20.7	24.8	13.6	18.2	22.8	27.4	14.9	19.9	24.9	29.8	

[a] After H. E. Korab, "Schemes for dissolving caustic before addition to bottle washer," Am. Bot. Carbonated Bev., Washington, 1956.
[b] Ascertain either from the catalog description of your soaker or, and preferably, by actual measurement the number of gallons each caustic tank holds up to a fixed mark on each gage glass or any other type marker indicating solution level. If the volume is given in cubic feet, multiply by 7.5 to convert into gallons. One gallon of water weighs 8.3 pounds.
[c] Determine either from the label on the barrel or by the manufacturers' statement the percentage of caustic content in the particular compound used.
[d] Prepared by dissolving 400 pounds of dry caustic in 70 gallons of water.

stock solution may be estimated as detailed in Chapter 15 or with A.B.C.B. caustic test tablets. It is necessary, however, to dilute the stock solution before making the test. The simplest procedure is to mix 10 milliliters of the concentrated solution with 90 milliliters of water and to run the A.B.C.B. caustic test in the customary manner. A value of 5 per cent sodium hydroxide in the test solution means that the stock solution contained 50 per cent caustic.

Other detergents, such as trisodium phosphate or proprietary products can be dissolved in the caustic soda solution if desired. For instance, if 15 per cent of trisodium phosphate per pound of caustic soda is to be used, then, in the example given, 50 pounds of trisodium phosphate should be added with the 400 pounds of caustic soda.

In order to pump all or a portion of the concentrated caustic solution into any compartment of the soaker, close valve D and open valve I.

When adding predissolved caustic soda solution to the washer by such a method, two men are required, one at the supply tank where the motor control switch is located and the other on top of the soaker to feed the concentrated stock caustic solution into the proper washer tank.

The system can be modified so that it requires the services of only one man. Such a modification is shown in Figure 18. Here two motor control switches are used and the sight glass on auxiliary tank L is placed in such a position that it can be observed from the motor switch located near the mixing tank A.

The process of predissolving the caustic soda in tank A is identical with the previously described procedure. After the caustic soda has dissolved, a portion of the concentrated solution formed is pumped into tank L. An empty caustic soda drum, equipped with a gage glass graduated in 2-gallon divisions, is suitable for this purpose. The stock caustic solution is fed from this auxiliary tank through a hose to any compartment of the soaker by gravity. The hose may be replaced by a permanent pipe manifold distribution system having an iron gate valve over each compartment. As a safety measure against improper operation of the motor control switch, an overflow pipe N should be installed as indicated on the auxiliary measuring tank.

This system is particularly advantageous for bottlers who use paper labeled bottles, for the alkali solutions in such instances

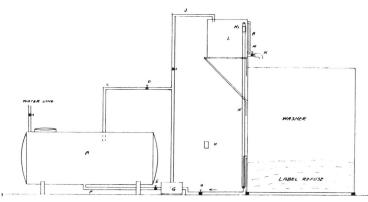

FIGURE 18. Caustic soda predissolving system modified for one-man operation (Courtesy of the American Bottlers of Carbonated Beverages)

A—Mixing and storage tank
B—Gage glass, marked to indicate gallons
C—Flow return pipe
D, E, I, M, O—Iron gate valves
F—One-inch black-iron feed line to pump
G—Pump; iron-construction, open-type
H—Motor-control switch
H_1—Motor-control switch—optional
J—One-inch black-iron feed line to washer
K—Hose of proper length
L—Measuring tank
N—Overflow

have to be dumped more frequently because of the build-up of paper label refuse in the soaker. With this pumping system, the washing solution above the label refuse can be drawn off and after the refuse is discharged from the washer, the caustic soda solution can be returned to the soaker with sufficient additional stock caustic solution to bring the washing solution up to the desired caustic strength.

The number of times that a portion of the alkali solution can be reused varies with the size of the alkali tank, the number of bottles washed, and the amount of labels removed, and can only be determined by estimating the caustic concentration as well as the amount of dirt present from the color of the washing solution. Korab is of the opinion that the caustic solution drawn from above the label refuse can be reused at least twice before it becomes necessary to dump the entire washing solution completely.

e. Label Removal

Removal of labels from beverage bottles is an important part of bottle washing. In the carbonated-beverage industry this is done during the washing step. The caustic soda solution of the washer attacks the glue and cellulose of the paper label or the adhesive and aluminum of the aluminum foil label and the label is thus able to drop from or be removed from the bottle. In general, whether "hydro" or immersion soaker type washing machines are used, the outer label coating must be attacked or wetted to a sufficient extent so that the washing solution can penetrate the paper and react with the adhesive. When the adhesive property is destroyed, the label can be removed from the bottle. Tenney [7] has suggested and discussed the use of alkyl aryl sulfonate wetting agents, such as Nacconol NR, that is a sodium alkyl aryl sulfonate, in concentrations of the order of 0.18 per cent to assist in label removal.

The removal of aluminum foil paper backed labels is a more difficult task than the removal of paper labels. Cooper [5] has discussed this problem in detail. He points out that to remove such labels, all or nearly all of the aluminum must be dissolved before the soaker solution can attack the paper and adhesive. Sodium hydroxide is used up in dissolving the aluminum. Indeed, chemically we state that 3 moles of sodium hydroxide react with 1 mole, or molecular weight, of aluminum:

$$2Al + 6NaOH \rightarrow 2NaAlO_2 \cdot Na_2O + 3H_2$$

Thus sodium hydroxide is consumed in dissolving the aluminum with the formation of sodium aluminate. When the free caustic is exhausted in this process, the washing solution will not be able to remove additional labels and there will not be sufficient free causticity to sterilize the bottles being washed.

The simple, unmodified, tablet method for determining caustic strength cannot be used when bottles having aluminum foil labels are being washed because of the formation of sodium aluminate. The amount of caustic must be determined in the manner detailed in Chapter 15 for the estimation of sodium hydroxide causticity in the presence of aluminates.

Cooper quotes one company as stating, "the labels and foil on 100 cases of 12-ounce bottles contain approximately 1.96 pounds of metallic aluminum and require approximately 8.08 pounds of caustic for their removal." Thus 4 pounds of caustic soda are needed for every pound of aluminum.

Since additional caustic is required for reaction with the glue, dirt, and debris in the bottles, salts in the water, and to replace the carry-over adhering to the bottles as they move to the next compartment, only a limited number of cases of bottles with foil labels can be washed before the caustic will have to be replenished. As an example, Cooper cites the use of caustic soda in a Meyer Dumore four-compartment bottle cleaner No. 412. This machine has compartments of 712-gallon capacity, equivalent to 5,930 pounds of water. About 59 pounds of 76 per cent caustic soda are required for each percentage of caustic wanted. Since one hundred cases of aluminum foil labels on 12-ounce bottles require 8.08 pounds of caustic for removal, five hundred cases will require 40.5 pounds of caustic. Taking into consideration the caustic lost by reaction with dirt, glue, etc., five hundred cases of 12-ounce bottles will use up 45 to 50 pounds of caustic. If the soaker was charged with 3 per cent caustic solution, then approximately one third of this will have been used up by the five hundred cases. Therefore, it is essential that a constant check be made to see how much free caustic remains so that the requirements of various State regulations can be met. Cooper recommends the following procedure.

Determine the weight of caustic needed to bring the solution in each compartment up to the regular percentage carried. In the case of the four-compartment soaker that we have been using for illustration, this will be 3 per cent in compartments 1 and 2, and 1.5 per cent in compartment 3. Do not add more caustic soda at this time.

Obtain from the label manufacturer, or determine, the weight of aluminum in one hundred labels of the type being used. Calculate the number of bottles or cases of bottles that will be washed in a given period of time, such as from 8 a.m. to 12 noon, and from this number calculate the weight in pounds of aluminum that will be dissolved. Then for every pound of aluminum to be dissolved during the washing period selected, add 4 pounds of caustic soda to the estimated amounts in compartments 1 and 2. Do not add extra caustic soda to compartment 3.

Weigh out the total amount of caustic soda required for each compartment, dissolve in the usual manner, and add to the tanks. If a predissolving system is being used, calculate the number of gallons of concentrated caustic soda solution equivalent to the pounds of caustic soda required and add these volumes to the tanks.

Fill the fresh-water compartment and make certain that all the tanks are heated to the proper temperature; in this instance, compartment 1 at 130°F, compartment 2 at 160°F, and compartment

3 at 130°F. Note that these temperatures are higher than those given in Table 13-3. The prerinse, fresh water, and final rinses must be operated at temperatures that will not result in bottle breakage because of thermal shock. Empty, clean, and refill the fresh-water tank daily. In addition test the washed bottles frequently with indicator solutions, as detailed in Chapter 15.9. Cooper stresses that he recommends the use of only caustic soda and not of mixtures of caustic soda and phosphates or silicates. He adds, "caustic is the most efficient and cheapest chemical for both removing aluminum foil labels and sterilizing bottles."

After preparing the caustic soda solutions as detailed, take samples, analyze them, and keep a record of the percentage of caustic in the freshly charged compartments. After washing the calculated number of bottles or cases of bottles, retest the caustic solution in compartment 1 for free causticity. If it has fallen below the 3 per cent required, calculate the weight of caustic soda needed to bring it up to standard requirements and record this weight, but do not add any caustic soda until the following calculation is made. Calculate again the weight of aluminum to be dissolved in the succeeding period of time and the weight of the caustic soda required for this purpose. Add this weight to the weight found necessary to bring the caustic solution up to standard. Dissolve the total weight in the customary manner and add to the tank or add the equivalent volume of predissolved concentrated caustic solution.

f. Recovery of Caustic Solutions

If the beverage plant has facilities for settling and reclaiming tanks, it may be economically sound practice to recover used caustic solutions. Some mention of this has been made in section 13.3.d on predissolving of caustic soda. Most beverage manufacturers do not attempt to recover caustic solutions. It is their practice to use such solutions for a given time period and then discard them. Others draw off the used caustic solutions into special tanks, the caustic is recovered, the solution is fortified, and is then used again. The second procedure will result in a saving of washing chemical.

(1) Evaluation of Caustic Solutions for Recovery

The following method [5] has been suggested for evaluating the possibility of reusing a caustic solution. Calculate the weight of aluminum per case of bottles from the weight of aluminum per hundred labels and designate this as "unit weight."

Empty and wash compartment 3 of a four-compartment washer about twice as often as tanks 1 and 2. Add the number of gallons held by compartment 1 to those held by compartment 2 and multiply this sum by 0.1176. This will yield the number of pounds of aluminum that will be dissolved. Divide this by the weight of aluminum per case, that is the "unit weight" and the quotient will be the number of units of cases that may be washed before compartment 3 is to be cleaned and recharged with 1.5 per cent caustic soda solution.

Let G_1 = gallons in compartment 1
G_2 = gallons in compartment 2
then
$$(G_1 + G_2) \times 0.1176 = \text{pounds of aluminum, dissolved} \quad [1]$$
and
$$\frac{\text{pounds of aluminum dissolved}}{\text{unit weight}} = \text{cases to be washed} \quad [2]$$

For example, if compartments 1 and 2 each hold 712 gallons, then by substitution in [1]:

$$(712 + 712) \times 0.1176 = 167.46 \text{ lb of aluminum}$$

If the unit weight is 0.0196 pound of aluminum per case then by substitution in [2]:

$$\frac{167.46}{0.0196} = 8,554 \text{ units or cases}$$

Thus when 8,554 unit weights or cases have been washed, compartment 3 should be emptied, washed, and recharged.

To determine when compartments 1 and 2 should be cleaned and recharged, the factor 0.196 is used:

$$(G_1 + G_2) \times 0.196 = \text{pounds of aluminum dissolved} \quad [3]$$

In the example already cited, this is:

$$(712 + 712) \times 0.196 = 279.1 \text{ lb of aluminum}$$

Again by substitution in [2]:

$$\frac{279.1}{0.0196} = 14,240 \text{ units or cases}$$

That is, when 14,240 cases have been washed, the entire unit should be emptied, cleaned, and recharged. Once such a system has been calculated, instructions can be attached to the soaker giving the number of cases after which compartment 3 is to be emptied, cleaned, and recharged and again when the entire soaker is to be emptied, cleaned, and recharged.

In a five-compartment washer, tank 1 can be charged in the customary manner and compartments 2, 3, and 4 can be treated as if they were compartments 1, 2, and 3 of a four-compartment soaker. One must not forget to check tank 1 of the five-compartment washer in the usual way and it, too, must be brought up to proper strength, but it is not necessary to add additional caustic soda.

(2) Physical Recovery

When it is desired to recover the caustic soda solution, then the procedure outlined before is followed up to the point where the entire unit is to be emptied. Then solutions originally containing 1.5 per cent of caustic soda or less are permitted to run to the drain. Caustic solutions of higher strength are piped or run to a settling tank. One type of treatment is physical and is similar to that described in section 13.3.d; the other type of treatment is a chemical one.

(3) Chemical Recovery

In the chemical recovery of caustic use is made of the old process for the manufacture of sodium hydroxide before the electrolytic method became the important one. The old process depended on the reaction of a carbonate with slaked lime forming insoluble calcium carbonate and sodium hydroxide. The reactions are

$$\underset{\text{lime}}{CaO} + \underset{\text{water}}{H_2O} \rightarrow \underset{\text{slaked lime}}{Ca(OH)_2}$$

$$\underset{\substack{\text{sodium}\\\text{carbonate}}}{Na_2CO_3} + \underset{\substack{\text{slaked}\\\text{lime}}}{Ca(OH)_2} \rightarrow \underset{\substack{\text{caustic}\\\text{soda}}}{2NaOH} + \underset{\substack{\text{calcium}\\\text{carbonate}}}{CaCO_3}$$

The caustic soda in the soaker is depleted because of two principal reactions: (1) the carbon dioxide of the air is taken up rapidly by the caustic soda, forming sodium carbonate and at the same time reducing the caustic alkalinity and (2) the bottles being washed have adhering to the walls acid, sugar, color, and other components of the drink and labels, paper, glue, starch, etc., and refuse material. The acid will neutralize some alkali; some caustic will be used up in caramelization of the sugar; additional caustic will be used up in reactions with the labels, paper, etc. Even when the caustic is not being used for washing it will take up carbon dioxide from the air, forming sodium carbonate.

The apparent loss of caustic by conversion to carbonate, the loss of caustic by reaction with some acids, and some other losses attributable to some chemical reactions can be recovered, but the caustic lost by adhering to the bottles and that washed away in the rinse water cannot be recovered.

To recover the caustic converted to carbonate and to some salts by reaction with acids, such as citric or tartaric, the spent washing solution is drawn off into a tank where it is treated with lime and heat. The precipitates of calcium carbonate, calcium citrate, calcium tartrate, calcium sulfate, etc., that are formed are permitted to settle and to clarify the washing solution in the settling process in a manner analogous to that of clarifying water (see Chapter 5). After the solution has cleared, it may be pumped back to the soaker and brought up to proper strength in the manner detailed in the preceding sections.

The recovery tank may be of any shape but it must be large enough to hold all of the spent washing solution and still allow at least one foot above the liquid level to permit foaming and boiling without overflow when the lime is added. It should be placed preferably on the floor of the story immediately above the soaker so that a minimum of piping is required but not so close as to interfere with cleaning of the washer. It must be equipped with an outlet port above the level at which the precipitate collects for drawing off the regenerated caustic and another outlet port at the bottom leading to the sewer for cleaning the recovery tank and disposal of the sediments. A steam line must be connected to a series of steam outlets in the tank or a pipe with holes can be used. An imput line must lead from the soaker to the recovery tank.

The concentration of free caustic and total alkali is determined by the A.B.C.B. tablet test. The spent soaker solution is transferred by a pump, similar to the type used in "Hydro" washers to the recovery tank by closing the return valve to the washer and opening the valve leading to the recovery tank. The spent solution is brought to 180°F by injecting steam. The proper amount of lime is added according to Table 13-7, the temperature is raised to 200°F with steam and is held at 200° to 217°F for 30 minutes, after which the steam is turned off. The reaction mixture is allowed to settle for about 2 hours which is generally sufficient for full precipitation and clarification after which the regenerated and clarified washing solution may be returned to the soaker by keeping the valve to the sewer closed and opening the valve in the line connecting the tank to the soaker. The washing solution is then brought up to proper strength by the method detailed.

The economies to be gained by using this process are the saving in costs of caustic over lime (which are appreciable in times of caustic stringency) and the heat required for the regeneration. An additional economy is the reduction in time and labor required for frequent dumping of spent soaker solution. One must include, however, the labor involved in the regeneration process, although some of this is counterbalanced by the costs saved in dumping and recharging the soaker.

TABLE 13-7

Amounts of Lime to Be Used for Caustic Recovery *

Number of tablets used in A.B.C.B caustic test	Number of tablets used in A.B.C.B. alkali test							
	1	1½	2	2½	3	3½	4	4½
0	5.8	8.7	11.6	12.4	13.1	13.8	15.3	17.4
1	2.9	5.8	8.7	9.4	10.5	11.6	12.7	13.8
2	—	2.9	5.8	6.9	8.0	8.7	10.2	12.4
3	—	—	2.9	3.6	4.2	5.8	6.6	8.7
4	—	—	—	—	—	2.9	4.2	5.8

* Pounds of lime to use per 100 gallons of soaker solution.

LITERATURE CITED

1. J. H. Buchanan, in *Bottlers' and Glass Packers' Handbook*, p. 51, New York, McGraw-Hill.
2. A. E. Abrahamson, in Morris B. Jacobs, Ed., *Chemistry and Technology of Food and Foods Products*, 2nd Ed., Vol. I, Chapter XVI, New York, Interscience Publishers, 1951.
3. J. F. Hale, in M. B. Jacobs, Ed., *Chemistry and Technology of Food and Food Products*, 1st Ed., Vol. II, Chapter XXIII, New York, Interscience Publishers, 1944.

4. J. H. Buchanan, M. Levine, and C. E. McKelvey, in *Bottlers and Beverage Manufacturers Universal Encyclopedia*, Chicago, Exposition Company of America, 1925.
5. W. C. Cooper, "Efficient Bottle Washing," Bull. No. B 139, Cudahy, Wis., Geo. J. Meyer Manufacturing Co., 1957.
6. H. E. Korab, "Predissolved Caustic," Washington, Am. Bottlers Carbonated Beverages.
7. R. I. Tenney, In *Bottlers' and Glass Packers' Handbook*, p. 20, New York, McGraw-Hill.

SELECTED BIBLIOGRAPHY

Plant Operation Manual, Washington, Am. Bottlers Carbonated Beverages, 1940.

H. E. Korab, *Technical Problems of Bottled Carbonated Beverages*, Washington, Am. Bottlers Carbonated Beverages, 1950.

H. E. Medbury, *The Manufacture of Bottled Carbonated Beverages*, Washington, Am. Bottlers Carbonated Beverages, 1945.

M. B. Jacobs, Ed., *Chemistry and Technology of Food and Food Products*, New York, Interscience Publishers, 1951.

R. H. Morgan, *Beverage Manufacture*, London, Attwood & Co., 1938.

W. C. Cooper, "Efficient Bottle Washing," Bull. No. B 139, Cudahy, Wis., Geo. J. Meyer Manufacturing Co., 1957.

Chapter 14

SPOILAGE

Spoilage of carbonated beverages, or for that matter of any food or food product, may be defined as any change from a desirable condition, grade, or standard of the product. These changes may result from physical, chemical, biochemical, or microbiological causes or action. The entire manufacturing process in the preceding chapters has been detailed from the point of view of avoiding spoilage entirely or of keeping it to a minimum. The causes of spoilage and its control will be considered in greater detail here.

From a manufacturing outlook, one must realize that spoilage may affect not only the finished drink but the raw materials as well. Very large losses may result from the spoilage of raw materials and once these spoil, it may be impossible to recover the raw material or make beverages that will not spoil.

As stressed in the preceding chapter, good sanitation practices and cleanliness are paramount in avoiding spoilage losses.

1. PHYSICAL SPOILAGE

The deleterious effects of light and heat on carbonated soft drinks are considered physical spoilage. Other types of physical spoilage are poor appearance resulting from the presence of precipitates or extraneous matter.

a. Light

Hale [1] pointed out that the bad effects of sunlight on beverages is known to everyone who has been in the business of manufacturing soft drinks for any length of time. The spoilage effects of light can be considered in two main categories: (a) undesirable flavor changes, characterized as terpenelike, oily, sour, etc. and (b) offtastes and odors in beverages made with true fruit bases. Terpenelike and oily tastes are found principally in citrus-flavored beverages. Light-induced rancidity occurs in products containing fixed oils, like vegetable oils including brominated vegetable oils.

In the second category is spoilage attributable to the fading of colors. Not all colors used in soft drinks are light fast and even colors with good light fastness will fade when subjected to strong sunlight. To prevent spoilage from exposure to light, it is necessary to take every precaution to keep carbonated soft drinks out of the direct rays of the sun not only in the plant but also on the truck while being transported and in the store or other sales outlet. It is for this reason that bottlers are urged not to place fruit-flavored beverages, especially orange flavor, on top of delivery trucks.

b. Temperature Change

Wide variation in temperature may cause undesirable changes in soft drinks. High temperatures (of the order of 100°F) may affect flavor adversely. It is well to recall that carbon dioxide is far less soluble in water at high temperatures than at low temperatures. This will tend to cause leaks in bottled goods. Some cans may not be able to withstand the increase in pressure caused by increase in the ambient temperature (see Chapter 10.6).

Freezing temperatures may also have deleterious effects. Some colors and flavorings are much less soluble at very low temperatures. These may precipitate out of solution and may not go back into solution readily as the temperature is raised, or if they do go back in solution the shade of color or the flavor may be changed.

One thing to remember about freezing temperatures in the plant is that some of the flavoring substances in a flavor extract or flavor emulsion are far less soluble at very low temperatures than at normal temperatures. These may precipitate out of solution and, as in the case of beverages themselves, may not readily go back into solution. If a beverage is being prepared with such a precipitated flavor preparation, although the proper volume of extract or emulsion is added, this volume will not contain the proper amount of flavor. Therefore the finished beverage will not contain the proper amount of flavoring either and thus may be considered spoiled.

Particular care must be taken not to expose flavor emulsions to freezing temperatures, for freezing is a method for breaking some emulsions and this might well happen to the flavor emulsion.

c. **Appearance**

The appearance of a beverage is considered a physical property and, in this sense, if the appearance is substandard, the beverage is considered spoiled. The presence of rings in the necks of bottles, or of precipitates of any type, or of off-shades are all evidences of spoilage.

The appearance of a ring in the neck of a bottle of a cloudy beverage indicates that the emulsion of the beverage has broken down. This may, perhaps, be due to the beverage having been made with a flavor emulsion that had broken. "Ringing" may also result from improper or inadequate homogenization by the flavor manufacturer or during flavor manufacture. The oil globules should be reduced to less than 4 microns in size. Brownish or reddish-brown rings in the neck or elsewhere in bottles may be due to the precipitation of ferric hydroxide. Occasionally oil rings are formed as a result of oil from a solid carbon dioxide converter getting into a beverage.

Flocculent precipitates may be the result of poor water-treatment practices or poor filtration of the water or to microbiological activity or the precipitation of a coloring matter or its metal lake. Care must be taken to see that no brush bristles or other foreign material remains in the bottles after washing and rinsing.

Good manufacturing practice as recommended in this text will avoid spoilage of these types.

d. **Boiling and Gushing**

In a sense, boiling at the filler and gushing when the beverage is opened is a type of physical spoilage. This topic has been discussed in detail in Chapter 10, section 7 and need not be repeated here.

2. CHEMICAL SPOILAGE

Chemical spoilage of carbonated beverages may result from reactions of the beverage ingredients with oxygen of the air or from reactions among incompatible ingredients. Some of these undesirable reactions may be induced or accelerated by a physical cause, such as light or heat. For example, terpenelike odors or rancidity may result from the oxidation of terpenes or of fixed oils in the beverage by the oxygen of the air which is either dissolved in the beverage or is in the neck of the bottle or the head space of the can. Oxidation of terpenes and rancidity are not the only undesirable oxidative changes that may occur. Many flavorings, for instance, aldehydes, are sensitive to oxidation and when oxidized may spoil the beverage flavor.

Toulouse [2] has discussed a number of causes for excess air in bottled carbonated beverages. Such excess air may be present because of faulty filling technique or faulty carbonation of the water. If the snift, that is, the device for releasing to the atmosphere the air to be removed from the bottle, is not working properly, a considerable quantity of air may be compressed into the bottle. Some air is unavoidably retained in the bottle under the crown before it is placed in position. The water pumps on the filter or carbonator may draw in air or the water may be saturated with air before carbonation. This is particularly true if water is used directly from city mains. This can be overcome by use of a deaerator (see Chapter 5) or the water to be used can be stored in an open tank prior to carbonation. This practice permits the air to escape from the water before it is used.

In low-pressure filling, a process in which the entire bottle with its contents of sirup and carbonated water is open to the atmosphere for several seconds, the carbon dioxide which is escaping will displace a major portion of the air from the head space in the bottle. With good operating technique, less than 5 per cent of the head space gases will consist of air. In high-pressure filling, however, double or triple this volume of air will be retained, for the bottle is under pressures well above atmospheric at all times during the filling process.

In an examination of some 340 samples, Toulouse found the following range of air retention:

Air range	Samples	
%	No.	%
0–10	105	30.8
10–20	73	21.4
20–30	81	23.8
30–40	42	12.6
40–50	9	2.6
over 50	30	8.8

Such oxidative reactions are catalyzed by light and are accelerated by heat. These undesirable reactions can be avoided or minimized by deaeration of the water used for carbonation and by keeping the amount of air in the neck or head space at its lowest value. Spoilage attributable to reactions accelerated by heat can be kept to a minimum by proper storage temperatures.

The deleterious chemical effects of too much iron in the beverage water or too much copper in the water (if the beverage is to be canned) have been discussed in great detail in prior chapters and need not be considered again here. Mention must be made again of the adverse effects of excess chlorine in and high alkalinity of the treated water. These, too, have been discussed in considerable detail in Chapter 5. Care must be taken to see that rinsing after washing is adequate to remove all of the caustic and washing compounds. Otherwise such chemicals could cause spoilage of the final beverage.

Other adverse chemical reactions may be caused by low pH, that is, high acidity. Just the proper amount of acid should be used. Indeed, if too much acid is used not only will the product be spoiled, but money will be wasted in using excess acid.

Invert sugar, as mentioned, has a different taste and flavor than sucrose and this must be taken into consideration as a possible type of undesirable change. This may result from too long storage of the beverage.

While, in general, hardness in the beverage water will not affect a carbonated soft drink, too much hardness is undesirable and may cause spoilage. Thus, if the calcium concentration is very high and tartaric acid is used as the acidulant, a precipitate of calcium tartrate may occur. Again if hard water is being used to prepare a soft drink that is colored with a synthetic dye, the calcium or magnesium lake of the dye may be precipitated.

3. BIOCHEMICAL SPOILAGE

The principal form of biochemical spoilage to which carbonated beverages are subject is caused by enzymic activity. Enzymes are a class of colloidal, soluble organic substances elaborated by living organisms that catalyze specific reactions. Enzymes are simple proteins or conjugated proteins. The principal enzymes affecting soft drinks are the carbohydrases and lipases. The carbohydrases such as sucrase (invertase) and amylases catalyze the hydrolysis of carbohydrates. The lipases catalyze the hydrolysis of fats and esters.

At times enzymic activity occurs in conjunction with the presence and metabolism of microorganisms, but this is not always the case for enzymes that are produced by microorganisms are active even after the organisms are dead.

Enzymes may be present in the natural products used in the manufacture of carbonated beverages, such as fruit juices, fruit concentrates, gum arabic and other natural gums, etc. Such enzymes may cause undesirable changes in the raw materials themselves or in the final beverage. For instance, sucrase or amylase may accelerate the inversion of sucrose. This may be desirable at times and at other times may be totally undesirable.

Greater disadvantages may stem from the presence of lipases in natural products. These enzymes hydrolyze not only fats but also, at times, esters. As a consequence of their fat splitting action, they may cause rancidity and off-odors in beverages containing fats, fixed oils, or brominated fixed oils. Since many of the most important flavoring components, whether natural or synthetic, are esters, it can readily be seen that esterase enzymic hydrolysis would split such esters and reduce the flavor intensity or even cause off-flavors, off-odors, and off-tastes.

In addition to the presence of enzymes in natural products, it must be remembered that many enzymes are used in processing the raw materials utilized in the manufacture of soft drinks. For instance, pectinases are used to depectinize some true-fruit flavors. It is possible that a depectinized true-fruit flavor, which has also been treated to remove sugar, may be used for the flavoring of a dietetic soft drink to which pectin has been added as a bodying agent. If sufficient enzyme is left in the flavor, it may act on and hydrolyze the added pectin and thus destroy the bodying effect desired.

4. MICROBIOLOGICAL SPOILAGE

Undoubtedly spoilage attributable to the activity of microorganisms is the most serious problem of the beverage manufacturer.

Microorganisms are plant or animal organisms of microscopic or submicroscopic dimensions. Plant microorganisms are generally classified into three major groups, namely, fungi, bacteria, and algae. As far as beverage manufacture is concerned, the fungi are the most important group of spoilage organisms for the yeasts and molds belong to this major group. The protozoa are the most important of the animal microorganisms.

Actually while individual microorganisms are microscopic in size, they form colonies, clouds, precipitates, sediments, and scums which are visible to the naked eye and are the direct signs of spoilage.

a. Fungi

The fungi are simple plant organisms containing no chlorophyll and have a vegetative structure that is not differentiated into roots, stems, or leaves. The fungi are classified as *Eumycetes,* a subphylum of the *Thallophyta* and comprise the true fungi, that is, the *Phycomycetes* or algalike fungi, the *Ascomycetes* or sac fungi that include the yeasts and molds, the *Basidiomycetes* or club fungi, like the mushrooms, and the *Fungi Imperfecti* like the aspergilli. Sometimes the *Myxomycetes* or slime molds and the *Schizomycetes* or bacteria are classified with the fungi.

The spoilage caused by fungi will be considered under the major groups of yeasts and molds.

b. Yeasts

The term yeasts is the common name used to denote a broad group of fungal microorganisms. Most yeasts are unicellular and reproduce by budding and asexual spores. Some yeasts, however, reproduce by sexual spores. Yeasts are somewhat larger than bacteria. Their cell wall contains chitin and within the cell is a large vacuole. *Saccharomyces cerevisiae* is a typical example.

Yeasts grow more readily in sirups, beverage acid solution (the optimum pH for yeast growth is 4.0 to 4.5), some flavors, and other beverage raw materials than any other class of microorganisms, under the conditions customarily existing during the manufacture of carbonated and uncarbonated soft drinks. Consequently, it is of the greatest importance for the beverage manufacturer to know the details of their occurrence and growth, and the factors controlling their growth. Hale [1] has discussed this subject. He pointed out that yeast growth is responsible for over 90 per cent of the spoilage in bottled carbonated beverages attributable to microorganisms.

Yeasts get into the beverage manufacturing process by way of: (1) raw-material containers, such as sugar sacks; (2) contaminated flavors and coloring matters; (3) dust and other air-borne material or dirt picked up during the manufacturing; (4) unsanitary plant operations; (5) contaminated equipment and pipe lines; (6) improperly or incompletely sterilized bottles or cans; (7) dusty crowns.

Hale made a study of the effect of the four major components in carbonated beverage manufacture, namely, sugar, acid, carbon dioxide, and water, on the growth of yeasts. The particular strain he used for his studies was isolated from spoiled ginger ale and simple sirup.

(1) Effect of Sugar

In Table 14-1 the rate of growth of yeast in sirups of different sugar concentration is tabulated. It can be seen that the growth of yeast in dilute sugar solutions, containing 1 per cent of sucrose, is very rapid. In 48 hours, the initial number more than doubles and in 5 days reaches 2,300,000 organisms. The yeasts grow faster in 5 per cent than in 1 per cent sugar solutions. Solutions of this sugar concentration, are not customary for the manufacture of the soft drinks themselves, but they are of the order of magnitude of sugar solutions that may occur in the siruper or in the lines leading to the siruper. It is clear why every effort must be made to clean such lines properly and thus avoid infection by yeasts.

TABLE 14-1

Rates of Yeast Growth in Sugar Solutions[a]
(Incubated at 28°C, solutions made in tap water)

Incubation time, hours	Yeast per ml in			
	1% sucrose	5% sucrose	27°Bé sirup (49.5%)	32°Bé sirup (59.1%)
Initial infection	127,000	120,000	190,000	205,000
8	143,000	195,000	261,000	782,000
30	192,000	580,000	941,000	1,680,000
48	290,000	940,000	3,600,000	2,150,000
60	385,000	2,110,000	7,410,000	2,895,000
85	870,000	11,700,000	8,600,000	3,752,000
119	1,600,000	26,500,000	9,100,000	4,492,000
145	2,300,000	43,800,000	9,900,000	6,631,000
162	2,650,000	57,300,000	12,600,000	7,389,000

[a] After J. F. Hale in Morris B. Jacobs, Ed., *The Chemistry and Technology of Food and Food Products*, 1st Ed., Vol. II, page 730, Interscience Publishers, New York, 1944.

It can be seen from Table 14-1 that yeasts still grow rapidly in 27° and 32° Bé sirups in which the sugar concentrations are 49.5 and 59.1 per cent respectively. As stressed in Chapter 2, the storage of sirups of less than 36° Bé, to which acid has not been added, is inadvisable for yeasts can grow in such sirups. Certain species of yeasts may grow even in sirups of 36° Bé or higher density so that the need for sanitary handling cannot be avoided merely by the practice of using high-density sirups.

(2) Effect of Acid

When acid is present, even an organic acid in such low concentration as 0.017 per cent equivalent to 10 grains per gallon, without sugar, yeasts do not grow as can be seen from Table 14-2. The addition of even 1 per cent sucrose changes the environment so that yeasts grow readily. Indeed, there is little difference from the rate of growth in a 1 per cent sucrose solution without any acid. With, however, 5 per cent sucrose solution, the same amount of acid, namely 0.017 per cent does reduce the growth. If the concentration of acid is increased to 0.085 per cent (50 grains per gallon), the yeasts die rapidly. They die even more rapidly when the acid concentration is raised to 0.34 per cent (200 grains per gallon). Insalata, however, found that heavy inoculations of yeast cells in sucrose solutions were not inhibited by 0.291 per cent of citric acid (171 grains per gallon).

TABLE 14-2

Rates of Yeast Growth in Sucrose-Citric Acid Solutions [a]
(Incubated at 28°C, solutions made in tap water)

Incubation time, hours	Yeast per ml in						
	10 gr citric acid/gal	1% sucrose 10 gr citric acid/gal	5% sucrose 10 gr citric acid/gal	5% sucrose 50 gr citric acid/gal	5% sucrose 200 gr citric acid/gal	10% sucrose 10 gr citric acid/gal	10% sucrose 50 gr citric acid/gal
Initial infection	139,000	119,000	135,000	150,000	140,000	119,000	140,000
8	129,000	121,000	......	......	......	......	......
30	93,000	133,000	169,000	47,000	6,100	159,000	......
45	4,000	140,000	195,000	......	......	......	......
60	200	162,000	220,000	......	......	......	......
85	0	197,000	233,000	11,200	......	......	......
119	......	257,000	251,000	......	......	240,000	......
145	......	770,000	292,000	400	320	......	200
162	......	1,200,000	305,000	50	......	......	......
180	......	2,700,000	310,000	......	......	2,120,000	......
220	......	2,509,000	295,000	......	15	......	......
260	......	2,000,000	290,000	......	......	1,900,000	......

NOTE: 10 grains of citric acid per gallon in 0.017% acid.
50 grains of citric acid per gallon in 0.085% acid.
200 grains of citric acid per gallon in 0.340% acid.
1 grain per gallon equals 0.0171 g. per liter.

[a] After J. F. Hale in Morris B. Jacobs, Ed., *The Chemistry and Technology of Food and Food Products*, 1st Ed., Vol. II, page 730, Interscience Publishers, New York, 1944.

Table 2-2 shows that sterility can be obtained in relatively dilute sirups, such as 24°Bé after 3 to 5 days; with 4 ounces of 50 per cent citric acid per gallon of sirup (approximately 850 grains per gallon) for the three-day period, with 2 ounces (approximately 420 grains per gallon) for the four-day period, and with 1 ounce (approximately 220 grains per gallon) for the five-day period.

It must be stressed that yeasts vary in their ability to become acclimated to different environmental conditions and thus the data presented cannot be considered as absolute.

(3) Effect of Carbon Dioxide

It can be seen by inspection of Table 14-3 that carbon dioxide has some inhibiting effect on the growth of yeasts. A comparison of the growths of yeast in 10 per cent sucrose solution containing 0.017 per cent of citric acid with and without 3.5 volumes of carbon dioxide shows that there is less growth when the carbon dioxide is present.

TABLE 14-3

Rates of Yeast Growth in Carbon Dioxide–Citric Acid–Sucrose Solutions [a]

(Incubated at 28°C, solutions made up in tap water)

Incubation time, hours	Yeast per milliliter in			
	1% sucrose, 3.5 volumes CO_2	5% sucrose, 3.5 volumes CO_2	10 gr citric acid/gal, 3.5 volumes CO_2	10 gr citric acid/gal, 10% sucrose, 3.5 volumes CO_2
Initial infection	135,000	131,000	142,000	121,000
8	139,000	162,000	125,000	......
30	142,000	510,000	79,000	......
45	161,000	......	7,000	139,000
60	187,000	......	500	......
85	263,000	9,600,000	10	......
119	730,000	23,000,000	......	195,000
145	1,240,000	47,000,000	0	......
162	1,710,000	58,000,000	......	......
180	2,100,000	75,000,000	......	340,000
220	1,900,000	93,000,000	......	......
260	1,800,000	94,000,000	......	430,000

[a] After J. F. Hale in Morris B. Jacobs, Ed., *The Chemistry and Technology of Food and Food Products*, 1st Ed., Vol. II, page 731, Interscience Publishers, New York, 1944.

Insalata [3] also studied the effect of carbon dioxide on yeasts. He performed his experiments with a strain of *Saccharomyces cerevisiae*, the ordinary yeast of the distiller and brewer. When 800 of such organisms were inoculated into each of twenty four 7-ounce bottles of carbonated water with a pressure of 5 volumes and when these bottles were incubated at room temperature, sterility was obtained at the end of 48 hours. Such samples had no sugar and consequently were not a favorable medium for yeast growth.

In his tests with prepared samples of beverages containing either 12 per cent of sugar (12° Brix) or 8.5 per cent of sugar (8.5°

Brix), having a pH of 4.5, carbonated at 3, 4, and 5 volumes of carbon dioxide, and each sample inoculated with 150 yeast cells, Insalata found three groups of results.

With 5 volumes of carbon dioxide, sterility was obtained on the third day in 8.5 per cent sugar and on the tenth day in 12 per cent sugar samples.

With 4 volumes of carbon dioxide, sterility was obtained on the eighth day in 8.5 per cent sugar and on the thirteenth day in 12 per cent sugar samples.

With 3 volumes of carbon dioxide, sterility was obtained on the 17th day in 8.5 per cent sugar and on the 20th day in 12 per cent sugar samples.

In each case sterility was obtained sooner in the 8.5 per cent sugar samples than in the 12 per cent sugar samples. Insalata thought that this might be due to the fact that the 12 per cent sugar solution was a better medium for yeast growth.

He stressed that the use of high carbon dioxide pressure was no substitute for cleanliness and good sanitary practices.

c. **Molds**

Molds are filamentous fungi composed of many cells, usually tubular or cylindrical, forming hyphae. The branched hyphae make up a mass of threads known as a mycelium. The aerial hyphae bear spores in sporangia or have free spores which are sometimes colored and give the mold a characteristic appearance. Molds lack chlorophyll, as mentioned, and consequently are saprophytic in their nutritional requirements. They need air for growth and can grow in and prefer acid media. Insalata found that high carbon dioxide pressures produce an environment unfavorable for mold growth.

Properly carbonated beverages usually do not have sufficient air to support the growth of molds. Because, however, molds can tolerate acids, they can grow on citric acid solutions and on acidulated sirups. They can also grow on the corks of crowns, on paper and cardboard boxes, on bags, wood, and consequently can cause a considerable amount of spoilage of raw materials used for the beverages and materials used in the beverage plant.

If molds do get into a beverage, they produce a characteristic musty off-taste and off-odor. If the plant is not properly sanitized and molds grow in the beverage plant, itself, the air and plant will have a musty and unpleasant odor. The sanitary practices stressed throughout the text will avoid this type of spoilage.

d. **Bacteria**

Bacteria are unicellular microscopic organisms that are classified with the plant kingdom as class *Schizomycetes* of the phylum *Thallophyta*. They are widespread, being found almost everywhere in the water, air, soil, food, and on the skin. Bacteria reproduce by binary fission. They may be coccoid or round in shape, bacilli or rods, or spirillum or screwlike. Some of them are motile.

When they grow in beverages and cause spoilage, their growth is made evident by the off-odors, off-tastes, and off-colors they cause. Some bacteria are autotrophic, that is, they are able to obtain energy from inorganic sources of nitrogen and carbon, or in some instances from sunlight. Other bacteria are heterotrophic, that is, they depend on organic substances for their nutritional needs. Some bacteria grow without access to oxygen or under low oxygen pressure and would find carbonated beverages a suitable environment if it were not for the fact that most bacteria cannot grow in acid media. There are, however, bacteria, known as acidophilic or aciduric bacteria that can grow at a pH of 3.0 to 4.5.

An organism that is considered characteristic of contamination with human intestinal wastes is *Escherichia coli*, Gram-negative, motile or nonmotile rods that appear in pairs or short chains. The presence of this organism in food or food products, water, or beverages of any type is considered an indication that such products or water are polluted with sewage. While *E. coli* is not ordinarily pathogenic, its presence is considered to indicate the possibility that pathogenic organisms, such as *Salmonella typhosa*, the organism causing typhoid fever, may also be present.

Hale points out that carbonated beverages have been found to have a low coliform group (to which *E. coli* belongs) index because of the germicidal effect of the combination of acidity and carbon dioxide concentration. The influence of the main ingredients of soft drinks on *E. coli* under experimental conditions is shown in Figure 19. Hale states that it is particularly significant that, in a solution of 3.5 volumes of carbon dioxide, 0.085 per cent of citric acid, and 10 per cent sugar, which is an environment simulating that of a bottled carbonated beverage, over 4,500,000 bacteria per milliliter were killed in less than 36 hours. The same trend is evident when there is no carbon dioxide, but the combined effect is even more certain.

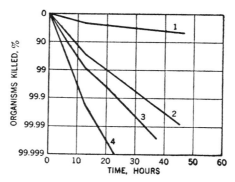

FIGURE 19. Graphic Representation of the Influence of the Principal Ingredients of Carbonated Beverages on *E. coli*

1—Tap water. 2—Tap water, 0.085% citric acid. 3—Tap water, 0.085% citric acid, 10% sucrose. 4—Tap water, 0.085% citric acid, 10% sucrose, 3.5 volumes carbon dioxide. Inoculum, 4,500,000 *E. coli* per milliliter (Unpublished data courtesy of W. A. Nolte, University of Maryland)

Insalata [3] restudied the effect of carbon dioxide on the growth of bacteria in carbonated beverages and concluded that high volumes of carbon dioxide produce sterility.

e. Algae

Algae is the common name of a heterogeneous group of plants capable of carrying on photosynthesis by means of the chlorophyll or other photosynthetic pigment they contain. They usually live in water, both fresh and marine, and on moist surfaces. Since they can manufacture carbohydrate, they do not live a saprophytic or parasitic life.

When conditions are favorable for their growth, that is, when they have water, light, carbon, and hydrogen, they grow rapidly. After accumulating in large masses, they die and give the water a disagreeable odor and taste. While such waters are not dangerous, they are undesirable and if proper water-treatment practices are not followed as detailed in Chapter 5, the algae, both live and dead, may get into a carbonated beverage and give it off-flavors, off-tastes, and off-odors. Ordinarily, except for the water, the carbonated beverage ingredients do not constitute a suitable medium for the growth of algae.

f. **Protozoa**

Protozoa is the common and classification name of a phylum of the animal kingdom comprising principally one-celled organisms of microscopic size and a few colonies. They use as their food organic matter, such as decomposed animal and vegetable wastes and also living animals and plants. If they are present in a water supply that is not properly treated, they may get into a carbonated beverage and spoil it by giving it an off-flavor and off-odor.

LITERATURE CITED

1. J. F. Hale, in M. B. Jacobs, Ed., *Chemistry and Technology of Food and Food Products*, 1st Ed., Vol. II, Chapter XXIII, New York, Interscience Publishers, 1944.
2. J. H. Toulouse, *Ind. Eng. Chem.*, **26**, 769 (1934).
3. N. F. Insalata, in *Bottlers' and Glass Packers' Handbook*, p. 59, New York, McGraw-Hill.

SELECTED BIBLIOGRAPHY

M. B. Jacobs, Ed., *Chemistry and Technology of Food and Food Products*, 2nd Ed., New York, Interscience Publishers, 1951.

Bottlers' and Glass Packers' Handbook, New York, McGraw-Hill.

M. B. Jacobs, *Chemical Analysis of Foods and Food Products*, 3rd Ed., Princeton, Van Nostrand, 1958.

Chapter 15

CHEMICAL ANALYSIS

There are two major aspects of the chemical and physicochemical analysis of carbonated nonalcoholic beverages. The first is concerned with the control of the manufacture of such beverages and the second with determining their composition both for investigative purposes and to ascertain whether or not they conform to certain standards, rules, and regulations.

It is beyond the scope of this text to treat certain types of composition analysis to any extent. For instance, it may be of interest to determine the dyes a given beverage contains. This type of analysis is considered in detail by Jacobs [1] and in *Methods of Analysis* of the Association of Official Agricultural Chemists.[2] It is seldom that the beverage manufacturer will find it necessary to do such a specialized type of analysis. This is equally true of the essential-oil ingredients. For such type of analysis, the reader is referred to Guenther.[3]

Throughout the text methods of analysis have been detailed in the descriptions of various raw materials, such as the sugars and the acids. Such methods assist the beverage manufacture in keeping strict control over his purchases. The methods, of course, need not be repeated in this chapter.

American Bottlers of Carbonated Beverages stress that proper control in the carbonated-beverage plant consists of keeping control of all amounts of all the ingredients used and regulating the manufacturing steps so that the finished product will always have the same quantities of sugar, or sweetening agent, acid, flavor, carbonation, and water, in a clean, suitable, and attractive container. A constant and standard predetermined volume of finished beverage should be present in each bottle or can.

The proper control measures that lead to such uniformity and quality control include accuracy in compounding the flavor and color additives, in measuring the amounts of sirups and acids, and in providing the volumes of carbonation. A wide variation in acid content may arise from a relatively slight inaccuracy in making up the stock acid solution. In a wholly analogous manner, errors in the preparation of the sirup or in the measurement and addition of the flavor and color result in nonuniformity in the final beverage and thus lack of quality control.

All of these factors can be controlled by careful attention to the details of the manufacturing process and by the use of tests in which relatively inexpensive, but exact measuring vessels, hydrometers, analytical balance, and other analytical equipment are used.

A number of bottlers' supply houses carry such testing equipment in stock. They are also readily obtained from chemical and scientific instrument supply firms that are located in every section of the country. Every beverage plant should have a complete testing laboratory and it is best to have at least two employees in the firm, trained to use such equipment and perform the tests.

The following simple methods will enable the manufacturer of carbonated soft drinks to determine (1) the character and quality of the water used, (2) the concentration of sugar in both sirup and final beverage, (3) the concentration of acid in both stock solution and in the finished beverage, (4) the volumes of carbonation, (5) the air content of carbonated beverages, (6) the concentration of caustic in bottle-washing compounds, (7) the efficiency of bottle rinsing, and (8) some miscellaneous tests.

1. WATER ANALYSIS

Standard methods for the analysis of water are detailed by the American Public Health Association.[4] Other methods are detailed by Furman,[5] by Betz [6] and by Solvay Engineering and Technical Service.[7]

a. **Total Solids**

In this method a known volume of water is evaporated to dryness in a weighed dish under standard conditions and the weight of the residue is determined after drying.

Procedure—Weigh accurately an empty, porcelain or equivalent evaporating dish on an analytical balance to the nearest milligram. Transfer with the aid of a 100-milliliter pipette 100 milliliters of the sample to the evaporating dish, or, for less accurate work, fill a 100-milliliter graduated cylinder to the mark with water and pour it into the evaporating dish and evaporate slowly on a water bath, steam bath, or in a drying oven held at 100°C. If a drying oven is used, be sure to regulate the temperature carefully and do not allow the sample to char. Dry the residue remaining after evaporation in a drying oven, thermostatically controlled at 103°C (217°F), for 1 hour. Remove from the oven, place in a desiccator to cool, and weigh. Subtract the weight of the empty dish from the weight of the dish plus the residue. The increase in weight, in milligrams, multiplied by 10, equals the total solids of the water in parts per million.

b. **Total Hardness**

The total hardness of water can be determined by either the soap method or by a method utilizing chelating compounds.

(1) Soap Method

The determination of hardness by the soap method is a rough approximation of the amount of calcium and magnesium in a water, for the method actually estimates the soap-consuming power of the water.

Procedure—Transfer 50 milliliters of the water being tested with the aid of a pipette or, for less accurate work, with a graduated cylinder to an 8-ounce bottle. Add a "standard" soap solution reagent in small portions at a time, shaking vigorously after each addition, until a marked permanent lather is obtained, which will remain for 5 minutes with the bottle laid on its side. A lather persisting for this length of time is indicative of the "true end point."

Before proceeding with the method, it is necessary to test the water for acidity. If it is acid, it must first be made neutral to methyl orange by the addition of 0.02 N ($N/50$) sodium carbonate solution.

(2) Chelation Method

Another method in use for the determination of hardness is based on the complexing and chelating action of ethylenediaminetetraacetic acid (EDTA) or related chelating compounds. This compound and its sodium salts have the ability to form chelate complexes with metal cations. A dye, known as Chrome Black T, also forms complexes with calcium and magnesium. In the pH range of 8.5 to 11.5, an aqueous solution of the dye, free from calcium and magnesium, is blue while the calcium and magnesium lakes are of burgundy color. Because the EDTA complexes are more stable than the dye compounds, the EDTA is able to react with the dye lakes liberating the free dye and chelating the calcium and magnesium. This turns the solution blue. The reagents are commercially available.

Procedure—Transfer 50 milliliters of the water sample to be tested with the aid of a pipette or with a graduated cylinder to a 250-milliliter Erlenmeyer flask. Add approximately 1 gram of the Chrome Black T-buffer-inhibitor reagent powder and titrate with standard EDTA solution, using a 25-milliliter burette. Multiply the number of milliliters of EDTA standard solution used in the titration by 20 to obtain the hardness as parts per million, expressed as calcium carbonate.

c. Total Alkalinity

The volume of 0.02 N standard acid solution required to neutralize a known volume of test solution is determined and the alkalinity is calculated.

(1) Methyl Orange Method

Procedure—Transfer 50 or 100 milliliters of the water to be tested with the aid of a pipette to an Erlenmeyer flask or a Phillips beaker or, for less accurate, work pour 50 or 100 milliliters of the water to be tested from a graduated cylinder to an Erlenmeyer flask or other vessel. Add 4 drops of phenolphthalein indicator solution. Add 0.02 N ($N/50$) standard sulfuric or hydrochloric acid from a burette graduated in 0.1 milliliters until the pink color disappears. The number of milliliters of acid used multiplied by

20, if a 50-milliliter sample was used, or by 10, if a 100-milliliter sample was used, gives the phenolphthalein alkalinity.

Add 2 drops of methyl orange indicator solution to the same sample and continue the titration until the color changes from yellowish orange to a reddish orange. The number of millititers of standard acid used multiplied by 20, if a 50-milliliter sample was used, or by 10, if a 100-milliliter sample was used gives the methyl orange or total alkalinity in terms of calcium carbonate.

(2) Methyl Purple Method

Methyl purple indicator solution may be used in place of methyl orange. The color change is from green to purple. An intermediate steel-gray tint provides a convenient warning of the approaching purple end point.

d. **Available Chlorine**

Chlorine determinations are required to ascertain if the levels of chlorination needed in water treatment systems utilizing chlorination (see Chapter 5) are being maintained and if the removal of all chlorine by activated carbon adsorbers is complete. Such methods may also be used to determine the concentration of hypochlorite sterilizing solutions.

The residual chlorine is determined by comparing the color of water samples to which *o*-tolidine (orthotolidine) reagent solution has been added with permanent color standards, representing known chlorine concentrations. *o*-Tolidine imparts a yellow or orange color to waters that contain a residual chlorine concentration.

Reagent—*o*-Tolidine Solution—Weigh out 1 gram of *o*-tolidine (melting point, 129°C), transfer to a 6-inch mortar and add 5 milliliters of hydrochloric acid (1 + 4). Grind the mixture to a thin paste and add 150 to 200 milliliters of water. The *o*-tolidine should go into solution immediately. Transfer to a 1-liter graduated cylinder and make up to 505 milliliters with water. Complete to the 1000-milliliter mark by adding hydrochloric acid (1 + 4). *o*-Tolidine reagent solution is relatively unstable. Therefore, it should be stored in amber bottles or otherwise be protected from direct sunlight and from contact with rubber. It should be replaced after 6 months. *o*-Tolidine reagent solution is commercially available.

Procedure—Fill the two or three tubes that the chlorine comparator apparatus, depending on the type, contains to the mark with the water to be tested. Add the specified amount of *o*-tolidine reagent solution to one of the tubes and mix. Place the tube in the comparator and place the other tube or tubes behind the color standards. Compare the color of the unknown with the standards and determine the matching color. Record the results as parts per million of chlorine.

When selecting the permanent standards, the range in which the comparator is to be used must be known. The directions as to the volume of water and *o*-tolidine to use may vary somewhat with different pieces of equipment, so that the instructions must be followed strictly to obtain correct results.

e. Total Iron

Ferric iron ions react with ammonium thiocyanate to form red ferric thiocyanate. The intensity of the color produced is proportional to the iron concentration. By treatment of water samples to convert all of the iron to the ferric state and by the use of prepared color standards, the total iron content of a water may be estimated. Such iron test kits are commercially available.

Procedure—Transfer 50 milliliters of the water to be tested with the aid of a pipette or with a graduated cylinder to a flask. Add 5 milliliters of hydrochloric acid (1 + 1) and 2 drops of 30 per cent hydrogen peroxide solution. Mix thoroughly and add 5 milliliters of ammonium thiocyanate reagent solution. Fill one of the comparison tubes to the mark with this solution and place it in the center compartment of the analyzer or comparator block.

To another 50-milliliter sample of the water, add 5 milliliters of hydrochloric acid (1 + 1), 2 drops of 30 per cent hydrogen peroxide solution, and 5 milliliters of distilled water. Transfer this mixture to the other two tubes of the comparator block and place them on each side of the test sample. Using the slide, compare the test sample with the permanent standards until the matching color is found. Report the iron concentration in parts per million.

2. ACID CONCENTRATION

The acids were discussed in Chapter 4 and the methods for evaluating their purity, strength, and assay were detailed. Usually in beverage manufacturing, the acid concentration of stock solutions is obtained by use of the hydrometer and that of beverages by titration with standard alkali solution.

a. Stock Solutions

(1) Hydrometer Method

A hydrometer or spindle is a float, usually cylindrical in shape, with a bulb in the middle, a slender upper stem, and a weighted bottom. The stem is graduated with a scale, usually in arbitrary units, such as Baumé, Brix, or Twaddell, etc., or directly in specific gravity. The Baumé scale is based on a modulus of 145, which has been most generally adopted, the relation between specific gravity at 20°C referred to 20°C and degrees Baumé for liquids heavier than water is:

$$°Bé = 145 - \frac{145}{\text{sp gr } 20°/20°C}$$

The following procedure is representative and is designed to ascertain the concentration of stock citric acid solution and thus assist in its control. If other acids are used in the preparation of a beverage, the tables given in Chapter 4 should be used.

Procedure—Transfer 7 to 8 fluid ounces of the stock citric acid solution into a 9-ounce graduated cylinder. Cool or warm to the temperature for which the hydrometer has been calibrated. Immerse the hydrometer in the solution. Push the hydrometer slightly below the equilibrium point and allow it to rise to its proper level. This is done to overcome surface tension and viscosity effects. When the hydrometer comes to rest and is free of air bubbles, read the scale number.

To read the hydrometer correctly, place the eye slightly below the surface level of the test solution. Raise it slowly until the surface, first seen as an ellipse, becomes a straight line. Take the reading at the position where this line cuts the hydrometer scale.

The concentration of citric acid solutions is usually expressed as a percentage. To convert from degrees Baumé to percentage of citric acid the following Table 15-1 can be used.

Table 15-1

Relation of Baumé Degrees to the Percentage of Citric Acid by Weight

(Reading given in percentage of citric acid crystals containing one molecule of water)

Table based on hydrometer calibrated for use at 77°F (25°C)

°Bé	Tenths of Degrees									
	0.0	0.1	0.2	0.3	0.4	0.5	0.6	0.7	0.8	0.9
20	39.6	39.8	40.0	40.2	40.4	40.6	40.8	41.0	41.2	41.4
21	41.6	41.8	42.0	42.2	42.5	42.7	42.9	43.1	43.2	43.4
22	43.7	43.9	44.1	44.3	44.5	44.7	44.9	45.1	45.3	45.5
23	45.8	46.0	46.1	46.3	46.5	46.7	46.9	47.1	47.3	47.5
24	47.7	47.9	48.2	48.4	48.6	48.8	49.0	49.2	49.5	49.7
25	49.9	50.1	50.3	50.5	50.7	50.9	51.1	51.3	51.5	51.7
26	51.9	52.1	52.3	52.5	52.7	53.0	53.2	53.5	53.7	53.9
27	54.1	54.3	54.5	54.7	54.9	55.1	55.3	55.5	55.7	55.9
28	56.1	56.3	56.5	56.7	56.9	57.1	57.4	57.6	57.8	58.0
29	58.2	58.4	58.6	58.8	59.0	59.2	59.4	59.6	59.9	60.1

TO USE: Assume it is desired to determine the citric acid content of pure citric acid solution which has been found to be 21.2°Bé. In line 21, read across until the column headed 0.2 is reached. The figure 42.0 means that a 21.2°Bé solution contains 42.0% citric acid. In a similar manner, it will be found that 25.6°Bé equals 51.1% citric acid; 28.0°Bé equals 56.1% citric acid, etc.

b. Acidity in Beverages

(1) Titrimetric Method

After the removal of the carbon dioxide from the beverage, a known volume of beverage is titrated with a standard alkali solution to a phenolphthalein end point. The concentration of acid can be calculated in grains per gallon or in percentage of acid. A representative method for citric acid is detailed. Standard alkali solution and titration kits are commercially available.

Procedure—Remove the carbon dioxide from the beverage by pouring it back and forth between two beakers, or by taking a known volume of beverage, measured in a graduated cylinder, transferring it to a beaker, boiling the beverage, cooling, returning the cooled beverage to the cylinder, and adding sufficient distilled water to make up for that lost by evaporation.

Withdraw a 10-milliliter aliquot (portion) with a pipette, taking care to keep the interference of bubbles of gas to a minimum. (If the sample was properly prepared, there should be no interference.) Place the 10-milliliter aliquot in a flask and bring to a boil to expel the remainder of the carbon dioxide gas if necessary. Add 3 drops of phenolphthalein indicator solution to the contents of the flask and titrate with $N/41$ sodium hydroxide solution or with 0.02 N ($N/50$) sodium hydroxide solution from a burette until a faint pink color appears. The appearance of the pink color indicates that all of the acid has been neutralized by the standard alkali solution. Compute the acidity by multiplying the number of milliliters of $N/41$ sodium hydroxide solution by 10 to give the acidity in grains per gallon as citric acid.

Example—If titrating 10 milliliters of a lemon soda by the given method showed that it took 6.4 milliliters of standard $N/41$ sodium hydroxide solution to neutralize the beverage, that is, to bring about the pink color change, the acidity was:

$$6.4 \times 10 = 64 \text{ grains of citric acid per gallon}$$

If $N/50$ sodium hydroxide solution is used, multiply by the factor 8.2 instead of 10, or if it is desired to express the acidity as percentage, the following equation may be used:

$$\frac{\text{ml NaOH} \times N}{1000} \times \frac{\text{equiv. wt. acid} \times 100}{\text{sample wt.}} = \text{per cent acid}$$

Example—If 10 milliliters of a sample of a carbonated beverage, acidified with citric acid (which has a molecular weight of 210 and since it is a tribasic acid, an equivalent weight of 70), required 8 milliliters of 0.02 N sodium hydroxide solution to produce a pink color, then:

$$\frac{8 \times 0.02 \times 70 \times 100}{1000 \times 10} = 0.112 \text{ per cent citric acid}$$

3. SIRUP DENSITY AND SUGAR CONCENTRATION

The concentration of the sugar in a sirup, or as it is commonly termed, the sirup density, is the basis of the concentration of sugar in the finished beverage. Consequently, sirup density must be determined as accurately as possible. This concentration can be estimated by a variety of methods, namely, by specific gravity, refractometric, polaroscopic, and chemical methods. The polaroscopic and chemical methods do not lend themselves readily for use by a manufacturer of beverages, unless he has a full laboratory and personnel available. The specific gravity (hydrometer) and refractometric methods of analysis are more suitable for his purposes. The methods used to determine the kind of sugar or kinds of sugar present in a sirup, that is, the amount of sucrose, invert sugar, D-glucose, fructose, etc., are beyond the scope of this book. For such methods and for the chemical and polaroscopic methods, the reader is referred to Jacobs [1] and to Browne and Zerban.[8]

a. **Sirups**

The concentration of sugar in a sirup can be determined readily by refractometric or specific-gravity measurements.

(1) Refractometric Determination

If a refractometer is available, the refractometric method is probably the simplest and most rapid way to determine the sugar concentration of a sirup.

The refractive index is a quantity which is constant for a pure substance under standard conditions of temperature and pressure. It is the ratio of the sine of the angle of incidence of a ray of light on the surface separating two media to the sine of its angle of refraction. The ray passing from a dense to a denser medium is bent toward the normal. Expressed mathematically:

$$\frac{\sin i}{\sin r} = n, \text{ the index of refraction}$$

where

$i = $ angle of incidence
$r = $ angle of refraction

With the Abbe refractometer and its modifications, the refractive index can be read directly; only a few drops of the sirup are needed, and either white or monochromatic (having one wave length) light can be used. This refractometer consists of a fixed telescope and two matched right-angle prisms. A ray of light passing through the prism and liquid and entering the prism at grazing incidence will emerge at less than a right angle for any value of n other than $n = \sin A$, where A is the acute angle in a triangle, one side of which represents the prism face through which the light emerges into the telescope. The prism must thus be rotated through an angle in order that the rays may be parallel to the telescope. This angle of emergence determines the refractive index. Rays striking the surface at an angle greater than the angle of emergence are totally reflected. By adjusting the light and dark portions of the field so that the line of demarcation is sharp and coincides with the cross hairs, the refractive index can be read directly on the scale. The prism box has a jacket so that the temperature can be controlled and the telescope has a set of Amici prisms for compensating for light aberration.

Procedure—Set up a 2-gallon aspirator bottle, partially fill it with water, and adjust the temperature of the water to 20°C. Connect the bottle by a rubber tube to the jacket of the refractometer. Allow the water to run through the refractometer jacket until the refractometer thermometer shows that the temperature is 20°C. Open the prisms, place two drops of sirup on the lower prism, close the prisms, sight through the telescope, and adjust the light and dark portions of the field by moving the knurled knobs so that the line of demarcation is sharp and coincides with the cross hairs. Read the refractive index on the scale with the aid of the magnifier of the instrument and refer to Table 2-6 for the corresponding sugar percentage, Brix value, and specific gravity.

(2) Hydrometer Method (Baumé Scale)

The use of a hydrometer spindle and the Baumé scale were explained in section 2.a.(1) of this chapter. The Baumé hydrometer is calibrated so that it can be read at a specific temperature, in the case of sirup hydrometers, at either 60°F or 77°F. If the temperature is higher or lower than that for which the hydrometer was calibrated, adjust the temperature of the sirup or use an appropriate temperature correction factor.

Procedure—Pour the sirup to be tested into a cylinder and allow it to stand long enough for the air to escape. Adjust the temperature to that at which the hydrometer was calibrated. Insert the Baumé hydrometer and allow it to sink slowly and find its own level. Be sure that it does not touch the sides of the cylinder and use care in placing the hydrometer into the sirup so that it does not submerge below the reading point. This avoids too long a wait for equilibrium to be established because of the high viscosity and surface tension of the sirup. Read the degrees Baumé at the meniscus. Note the temperature and correct for the temperature differ-

CHEMICAL ANALYSIS

ence from the standard temperature of the hydrometer by the use of Table 15-2.

TABLE 15-2

Correction Table for Baumé Sugar Hydrometer [a]

Temperature °F	Baumé hydrometer calibrated at 60°F	Baumé hydrometer calibrated at 77°F
50	subtract 0.2	subtract 0.6
55	subtract 0.1	subtract 0.5
60	no correction	subtract 0.4
65	add 0.1	subtract 0.3
70	add 0.2	subtract 0.2
75	add 0.2	subtract 0.1
80	add 0.3	add 0.1
85	add 0.4	add 0.2
90	add 0.5	add 0.3

[a] After American Bottlers of Carbonated Beverages. The temperature corrections listed in this table are only approximate. For accurate reading, the sirup temperature should be adjusted to the temperature for which the hydrometer is calibrated.

b. Liquid-Sugar Density

The Baumé and specific gravity hydrometers commonly used in beverage laboratories are not convenient for determining the solids in liquid-sugar solutions. Special Baumé hydrometers are sometimes used for this work, but the use of a Brix hydrometer greatly simplifies the determination and yields more accurate results. A satisfactory determination of solids by means of a device for measuring the density of the liquid sugar, such as a hydrometer, cannot be made with any degree of accuracy on heavy liquid sugars, because of their high viscosity and the difficulty of eliminating air bubbles from the solution. More accurate analyses of such liquid sugar can be made by diluting the product with an equal weight of water and making the necessary corrections in the results for this dilution.

If a refractometer is available, the sugar concentration can be determined more easily by the refractometric method.

(1) Refractometer Method

Proceed with the method as detailed in section 3.a.(1) of this chapter under *procedure*.

(2) Hydrometer Method (Brix Scale)

Apparatus—The apparatus required for the determination of liquid-sugar density for concentrations ranging from 67 to 80 per cent total solids content comprises:
  1. A balance
  2. Brix hydrometer, calibrated at 20°C, having a range of 30° to 40°Brix, with a thermometer and a temperature-correction scale of 10° to 50°C. The Brix hydrometer is similar to the Balling hydrometer, that is, it is a saccharometer in which the scale has been recalculated, originally by Brix, to refer to the percentage by weight of sucrose in a solution at 20°C.
  3. Two 1-liter copper or nickel beakers of nearly the same weight or two counterbalanced glass beakers, that is, beakers of equal weight.

Procedure—Draw a 1-quart sample of the liquid sugar from the tank car or tank truck at the time of delivery. Dry the two copper, nickel, or counterbalanced glass beakers; place one on each balance pan; and balance them, if necessary, by means of the balance weights. Pour about 400 milliliters of the liquid sugar into the left-hand beaker. Pour cold distilled or tap water into the right-hand beaker until the pans are almost balanced. With the aid of a pipette, add water or withdraw water from the beaker containing the water until the beakers are exactly balanced. Mix the contents of both beakers thoroughly by pouring from one vessel to the other and by stirring with a rod. The liquid sugar and water should be poured from one beaker to the other at least ten times.

Dry a glass cylinder and rinse it twice with small portion of the diluted liquid sugar. Fill it nearly to the top with the diluted liquid sugar. If any air bubbles are present, allow them to rise to the top before proceeding with the determination or deaerate the diluted sample by putting the cylinder and its contents under a vacuum for a few minutes.

Lower the hydrometer carefully into the cylinder. Push the hydrometer slightly below the equilibrium point and allow it to rise to its proper level. Leave it in the liquid for a few minutes so that it can come to the same temperature as the liquid sugar. Exercise care to see that the instrument floats freely and does not touch the bottom or the walls of the cylinder. Read the scale at the level of the surface of the liquid and not at the level of the film of liquid drawn up around the stem by capillarity. (Refer to 2.a.(1) of this chapter under *procedure.*) Multiply the result by 2 to obtain the sugar percentage.

It is desirable, for accurate results, to use the hydrometer at a temperature near to its calibration temperature of 20°C. It is best after preparing the diluted liquid sugar to adjust its temperature to 20°C. If, however, this cannot be done and the temperature of the liquid sugar varies from the temperature at which the hydrometer was calibrated, the reading must be corrected by applying the correction indicated on the scale beside the thermometer scale. The recommended practice is to read the Brix scale of the hydrometer first and then the temperature and the corresponding correction scales as the instrument is lifted out of the cylinder.

After the correct reading of the hydrometer has been determined, the Brix reading in percentage must be multiplied by 2 to get the correct solids content in Brix degrees of the original sample because the original sample was diluted with an equal weight of water.

c. Sugar Content of Finished Beverage

The concentration of sugar in a beverage can be determined by methods analogous to those detailed, either refractometric or specific gravity. The use of the Brix hydrometer is the more common practice.

Procedure—Remove the carbon dioxide from the beverage by pouring it back and forth between two glass beakers or containers, or take a known volume of the beverage, measured with a graduated cylinder, transfer it to a beaker, bring to a boil, cool, return the liquid to the same cylinder, and add distilled or tap water to make up for the water lost by evaporation during the boiling.

Transfer 7 to 8 ounces of the beverage from which the carbon dioxide has been removed to a 9-ounce cylinder. Adjust the temperature of the beverage to 20°C. Insert the Brix hydrometer and, taking the precautions detailed in sections 2.a.(1) and 3.b.(2) of this chapter, read the scale. If the temperature is higher or lower than that for which the hydrometer was calibrated, use an appropriate temperature correction factor from Table 15-3.

The sugar content of carbonated beverages cannot be measured with a hydrometer before gas removal because bubbles of gas form on the hydrometer and cause it to rise to such an extent that a correct reading cannot be obtained.

TABLE 15-3

Corrections for Brix Hydrometers [a]

Temperature °F	Brix hydrometer calibrated at 60°F	Brix hydrometer calibrated at 77°F
50	subtract 0.2	subtract 0.7
55	subtract 0.1	subtract 0.6
60	no correction	subtract 0.5
65	add 0.1	subtract 0.4
70	add 0.3	subtract 0.2
75	add 0.4	subtract 0.1
80	add 0.6	add 0.1
85	add 0.8	add 0.3
90	add 1.0	add 0.5

[a] After American Bottlers of Carbonated Beverages. The listed temperature corrections are only approximate. For accurate reading of the concentration of sugar in the beverage, the beverage temperature should be adjusted to the temperature for which the hydrometer is calibrated.

4. ARTIFICIAL SWEETENING AGENTS

The principal artificial sweetening agents used in carbonated beverages in the United States are saccharin and cyclamate. It is customary in manufacturing control to rely on the original weight of the sweetening agent added to the bottlers' concentrate as a measure of the amount in the final beverage. If a more accurate estimation is desired, the following relatively simple methods can be used.

a. **Saccharin**

(1) Taste Test

The presence of saccharin in a beverage can be detected by extracting it from the beverage and tasting the residue.

Procedure—Place 50 milliliters of the sample under test into a separatory funnel and acidify with hydrochloric acid. Extract with three portions of ethyl ether. Combine the ether extracts and wash once with 5 milliliters of water. Transfer the ether extract to an evaporating dish and allow the ether to evaporate spontaneously.

Taste the residue. Concentrations of the order of 1 milligram per 50 milliliters can be detected by taste.

(2) Salicylic Acid Test

The saccharin is converted to salicylic acid by fusion with sodium hydroxide and the acid is detected by means of the ferric chloride reaction.

Procedure—Transfer the residue, obtained as detailed in the taste test, to a nickel dish with the aid of ether and evaporate the ether. Add 10 milliliters of 10 per cent sodium hydroxide solution and evaporate to dryness. Bake for an hour on a hot plate and fuse cautiously over a Bunsen flame. Leach with water and transfer to a separatory funnel, make acid with hydrochloric acid, extract with ether, and wash as before. Transfer the ether layer to an evaporating dish, drive off the ether, and add a few drops of ferric chloride solution. A deep violet color is indicative of salicylic acid.

(3) Phenol–Sulfuric Acid Test

The following test is an official method of the Association of Official Agricultural Chemists.[2] It depends on the nitration of phenol-sulfuric acid with the nitric acid formed from the saccharin and the production of a color.

Procedure—Place 25 milliliters of the beverage sample into a separatory funnel. Acidify with 3 milliliters of hydrochloric acid and, if vanillin is present, extract with three 25-milliliter portions of petroleum ether, and discard. Extract the beverage successively with 50-, 25-, and 25-milliliter portions of an ethyl ether–petroleum ether mixture (1:1). Combine the extracts, wash once with 5 milliliters of water, and transfer to a 30-milliliter beaker. Allow to evaporate spontaneously and dry at room temperature.

Add 5 milliliters of phenol–sulfuric acid reagent, prepared by dissolving pure, colorless, crystalline phenol in an equal weight of sulfuric acid, and heat for 2 hours at 135° to 140°C. Cool, dissolve in a small quantity of hot water, and pour into about 250 milliliters of water. Add a small quantity of Filter-Cel, allow to stand for 3 hours or overnight, and filter. Add 10 per cent sodium hydroxide solution sufficient to make it alkaline and dilute to 500 milliliters. A magenta or reddish-purple color develops if saccharin is present. Yellow, buff, or pale-salmon shades are not considered significant.

(4) Determination

Saccharin can be estimated by converting the imino group to ammonia and determining the ammonia formed by nesslerization.

Procedure—Place 50 milliliters of the sample into a separatory funnel and add 2 milliliters of hydrochloric acid. Extract with two 50-milliliter portions of ethyl ether. Filter the ether extracts through a pledget of cotton into another separatory funnel and wash the combined extracts with 5 milliliters of water acidified with 1 drop of hydrochloric acid. Draw off the water layer, transfer the ether layer through the mouth of the separatory funnel to a beaker, and evaporate to dryness on a water bath. Add 5 milliliters of ammonia-free water and 6 milliliters of hydrochloric acid and evaporate, while constantly stirring, to about 1 milliliter on a hot plate. Repeat. Transfer to a 50-milliliter volumetric flask with the aid of ammonia-free water and make up to the mark. Transfer 2 milliliters with a pipette to a 25-milliliter Nessler tube, dilute to 25 milliliters, and add 1 milliliter of Nessler reagent. Compare the color produced with standards made with ammonium chloride, equivalent to 200 parts per million of the insoluble form of saccharin. Thus 0.2921 gram of ammonium chloride is equivalent to 1 gram of insoluble saccharin, $C_7H_5NO_3S$, and to 1.317 grams of sodium or soluble saccharin, $C_7H_4NNaO_3S$.

b. **Cyclamate**

(1) Sodium Nitrite Test

Sulfate can be split from either sodium cylcohexylsulfamate or calcium cyclohexylsulfamate by nitrite. Consequently, if a beverage is freed from sulfate ions by the use of barium chloride and on the addition of nitrite additional sulfate appears, the presence of cyclamate is indicated.

Procedure—Add 2 grams of barium chloride to 100 milliliters of the beverage being tested. Allow to stand for 5 minutes and filter or centrifuge. Acidify the filtrate or the supernatant liquid with 10 milliliters of hydrochloric acid and add 0.2 gram of sodium nitrite. A white precipitate of barium sulfate indicates the presence of cyclamate.

(2) Determination

The qualitative test described in the preceding section can be adapted for the quantitative determination of cyclamate in carbonated beverages.

Procedure—Transfer with the aid of a pipette a known volume of the sample being analyzed, containing about 100 milligrams of a salt of cyclohexylsulfamate, to a beaker. Dilute to approximately 200 milliliters with distilled water. Add 10 milliliters of concentrated hydrochloric acid and 10 milliliters of 10 per cent barium chloride solution. Mix thoroughly. If free sulfate is present as demonstrated by the formation of a precipitate of barium sulfate after adding the barium chloride solution, transfer the contents of the beaker to centrifuge tubes or a centrifuge bottle, making certain that the entire contents are quantitatively transferred, and centrifuge at a sufficient rate and time to pack the barium sulfate. Pour off the clear supernatant solution into another beaker. Add 10 milliliters of sodium nitrite solution and digest on a hot plate until the supernatant liquid is clear.

Prepare a Gooch crucible in the customary manner. Ignite, cool in a desiccator, and weigh. Filter off the precipitate on the asbestos pad of the Gooch crucible, wash, and dry in an oven at 100°C for 15 minutes. Ignite the crucible, place in a desiccator to cool, and reweigh. The difference in the weight of the tared prepared Gooch crucible and the crucible plus the precipitate is considered to be barium sulfate.

Each milligram of barium sulfate is equivalent to 0.9266 milligram of calcium cyclamate and to 0.8621 milligram of sodium cyclamate.

5. CARBONATION (GAS VOLUME) OF FINISHED BEVERAGE

In Chapter 10, the importance of proper carbonation was stressed and a brief description of the method of test was given. There is some controversy concerning the best method for testing the carbonation pressure in bottles of carbonated beverages. The following methods have been recommended by the American Bottlers of Carbonated Beverages Committee and is considered fair to all types of filling machinery by tests performed in the American Bottlers of Carbonated Beverages Laboratory.

a. **Beverages from Stock**

Procedure—Clamp the bottle in the frame of the tester. Pierce the crown but do not shake the bottle. Snift off the top gas quickly until the gage reading drops to zero. Make certain to close the valve the instant the needle touches zero. Shake the bottle vigorously until the gage gives a reading that additional shaking does not change. Record the pressure. Note and record the temperature. Obtain the volume of gas from Table 15-4.

b. **Beverages from the Bottling Line**

A beverage taken from the production line usually has little or no carbon dioxide pressure in the headspace. Consequently the determinations of gas volume of beverages taken from the production line are generally higher in pressure than those taken from stock. In order to correct for this difference, beverages taken from the production line should be allowed to remain quiet for at least 5 minutes prior to testing or the following procedure should be employed.

Procedure—Clamp the bottle in the frame of the gas volume tester. Puncture the crown and shake until maximum pressure is reached. Allow the bottle to remain quiet for 30 seconds in either an upright position or on its side. Snift the head pressure quickly to zero and then close the snift valve. Shake the bottle again until maximum pressure is reached. Record this pressure. Note the temperature and obtain the gas volume from the pressure and temperature values by use of Table 15-4.

6. GAS ANALYSIS

In a sense, the determination of gas volume detailed in the preceding section is a type of gas analysis. It is often necessary to make an actual analysis of the gases in a carbonated beverage and in the headspace above the beverage to determine the composition of the gases present. This information may solve spoilage and corrosion problems.

Such analyses can be carried out in the customary gas-analysis apparatus, such as an Orsat or more elaborate types of equipment

TABLE 15-4

Gas Volume Test Chart

(Showing volumes of carbon dioxide dissolved by 1 volume of water)

°F, in bottle	16[a]	18	20	22	24	26	28	30	32	34	36	38	40	42	44	46	48	50	52	54	56	58	60	62	64
45	2.7	2.9	3.1	3.3	3.4	3.6	3.8	4.0	4.1	4.3	4.5	4.7	4.8	5.0	5.2	5.4	5.6	5.7	5.9	6.1	6.2	6.4	6.6	6.6	6.9
46	2.8	2.8	3.0	3.2	3.3	3.5	3.7	3.9	4.0	4.2	4.4	4.6	4.7	4.9	5.1	5.3	5.4	5.6	5.8	6.0	6.1	6.3	6.4	6.6	6.8
47	2.6	2.8	2.9	3.1	3.3	3.5	3.6	3.8	4.0	4.2	4.3	4.5	4.6	4.8	5.0	5.2	5.3	5.5	5.7	5.9	6.0	6.2	6.3	6.5	6.7
48	2.6	2.7	2.9	3.1	3.2	3.4	3.6	3.7	3.9	4.1	4.2	4.4	4.6	4.7	4.9	5.1	5.2	5.4	5.6	5.7	5.9	6.1	6.2	6.4	6.6
49	2.5	2.7	2.8	3.0	3.2	3.3	3.5	3.7	3.8	4.0	4.1	4.3	4.5	4.6	4.8	5.0	5.1	5.3	5.5	5.6	5.8	6.0	6.1	6.3	6.4
50	2.5	2.6	2.8	2.9	3.1	3.3	3.4	3.6	3.7	3.9	4.1	4.2	4.4	4.5	4.7	4.9	5.0	5.2	5.4	5.5	5.7	5.9	6.0	6.2	6.3
51	2.4	2.6	2.7	2.9	3.1	3.2	3.4	3.5	3.7	3.8	4.0	4.2	4.3	4.5	4.6	4.8	5.0	5.1	5.3	5.4	5.6	5.7	5.9	6.1	6.2
52	2.4	2.5	2.7	2.8	3.0	3.1	3.3	3.5	3.6	3.8	3.9	4.1	4.2	4.4	4.5	4.7	4.9	5.0	5.2	5.3	5.5	5.6	5.8	5.9	6.1
53	2.3	2.5	2.6	2.8	2.9	3.1	3.3	3.4	3.6	3.7	3.9	4.0	4.2	4.3	4.5	4.6	4.8	4.9	5.1	5.2	5.4	5.5	5.7	5.9	6.0
54	2.3	2.4	2.6	2.7	2.9	3.0	3.2	3.3	3.5	3.6	3.8	3.9	4.1	4.2	4.4	4.5	4.7	4.8	5.0	5.1	5.3	5.4	5.6	5.7	5.9
55	2.2	2.4	2.5	2.7	2.8	3.0	3.1	3.3	3.4	3.6	3.7	3.8	4.0	4.1	4.3	4.4	4.6	4.7	4.9	5.0	5.1	5.3	5.4	5.6	5.8
56	2.2	2.3	2.5	2.6	2.8	2.9	3.1	3.2	3.4	3.5	3.6	3.8	3.9	4.1	4.2	4.3	4.5	4.6	4.8	4.9	5.0	5.2	5.3	5.5	5.7
57	2.2	2.3	2.4	2.6	2.7	2.9	3.0	3.2	3.3	3.4	3.6	3.7	3.8	4.0	4.1	4.3	4.4	4.5	4.7	4.8	4.9	5.1	5.2	5.4	5.6
58	2.1	2.2	2.4	2.6	2.7	2.8	3.0	3.1	3.3	3.4	3.5	3.6	3.8	3.9	4.0	4.2	4.3	4.4	4.6	4.7	4.8	5.0	5.1	5.3	5.5
59	2.1	2.2	2.4	2.5	2.7	2.8	2.9	3.1	3.2	3.3	3.4	3.6	3.7	3.9	4.0	4.1	4.2	4.4	4.5	4.6	4.8	4.9	5.1	5.2	5.4
60	2.0	2.2	2.3	2.5	2.6	2.7	2.9	3.0	3.1	3.3	3.4	3.5	3.6	3.8	3.9	4.0	4.2	4.3	4.4	4.5	4.7	4.8	5.0	5.1	5.3
61	2.0	2.1	2.3	2.4	2.6	2.7	2.8	2.9	3.1	3.2	3.3	3.5	3.6	3.7	3.8	4.0	4.1	4.2	4.3	4.5	4.6	4.7	4.9	5.0	5.2
62	2.0	2.1	2.2	2.4	2.5	2.6	2.8	2.9	3.0	3.1	3.3	3.4	3.5	3.6	3.8	3.9	4.0	4.2	4.3	4.4	4.5	4.7	4.8	4.9	5.1
63	1.9	2.1	2.2	2.3	2.5	2.6	2.7	2.8	3.0	3.1	3.2	3.3	3.5	3.6	3.7	3.8	3.9	4.1	4.2	4.3	4.4	4.6	4.7	4.8	5.0
64	1.9	2.0	2.2	2.3	2.4	2.6	2.7	2.8	2.9	3.1	3.2	3.3	3.4	3.5	3.7	3.8	3.9	4.0	4.2	4.3	4.4	4.5	4.7	4.8	4.9
65	1.9	2.0	2.1	2.3	2.4	2.5	2.6	2.8	2.9	3.0	3.1	3.2	3.4	3.5	3.6	3.7	3.8	3.9	4.1	4.2	4.3	4.4	4.6	4.7	4.8
66	1.8	2.0	2.1	2.2	2.3	2.5	2.6	2.7	2.8	3.0	3.1	3.2	3.3	3.4	3.5	3.7	3.8	3.9	4.0	4.1	4.3	4.4	4.5	4.6	4.8
67	1.8	1.9	2.1	2.2	2.3	2.4	2.6	2.7	2.8	2.9	3.0	3.1	3.2	3.4	3.5	3.6	3.7	3.8	3.9	4.1	4.2	4.3	4.4	4.6	4.7
68	1.8	1.9	2.0	2.2	2.3	2.4	2.5	2.6	2.7	2.9	3.0	3.1	3.2	3.3	3.4	3.6	3.7	3.8	3.9	4.0	4.2	4.3	4.4	4.5	4.6
69	1.7	1.9	2.0	2.1	2.2	2.4	2.5	2.6	2.7	2.8	2.9	3.1	3.2	3.3	3.4	3.5	3.6	3.7	3.8	3.9	4.1	4.2	4.3	4.4	4.5
70	1.7	1.8	2.0	2.1	2.2	2.3	2.4	2.5	2.7	2.8	2.9	3.0	3.1	3.2	3.3	3.4	3.6	3.7	3.8	3.9	4.0	4.1	4.2	4.3	4.4
71	1.7	1.8	1.9	2.1	2.2	2.3	2.4	2.5	2.6	2.7	2.8	3.0	3.1	3.2	3.3	3.4	3.5	3.6	3.7	3.9	4.0	4.1	4.2	4.3	4.4
72	1.7	1.8	1.9	2.0	2.1	2.3	2.4	2.5	2.6	2.7	2.8	2.9	3.0	3.1	3.2	3.3	3.5	3.6	3.7	3.8	3.9	4.0	4.1	4.2	4.3
73	1.6	1.8	1.9	2.0	2.1	2.2	2.3	2.4	2.5	2.6	2.8	2.9	3.0	3.1	3.2	3.3	3.4	3.5	3.6	3.7	3.8	3.9	4.0	4.1	4.2
74	1.6	1.7	1.9	2.0	2.1	2.2	2.3	2.4	2.5	2.6	2.7	2.8	2.9	3.0	3.1	3.2	3.3	3.4	3.6	3.7	3.8	3.9	4.0	4.1	4.2
75	1.6	1.7	1.8	1.9	2.0	2.2	2.3	2.4	2.5	2.6	2.7	2.8	2.9	3.0	3.1	3.2	3.3	3.4	3.5	3.6	3.7	3.8	3.9	4.0	4.1
76	1.6	1.7	1.8	1.9	2.0	2.1	2.2	2.3	2.4	2.5	2.6	2.7	2.8	2.9	3.0	3.1	3.2	3.3	3.5	3.6	3.7	3.8	3.9	4.0	4.1
77	1.6	1.7	1.8	1.9	2.0	2.1	2.2	2.3	2.4	2.5	2.6	2.7	2.8	2.9	3.0	3.1	3.2	3.3	3.4	3.5	3.6	3.7	3.8	3.9	4.0
78	1.5	1.6	1.7	1.9	2.0	2.1	2.2	2.3	2.4	2.5	2.5	2.6	2.7	2.8	2.9	3.0	3.1	3.3	3.4	3.5	3.6	3.7	3.8	3.9	4.0
79	1.5	1.6	1.7	1.8	1.9	2.0	2.1	2.2	2.3	2.4	2.5	2.6	2.7	2.8	2.9	3.0	3.1	3.2	3.3	3.4	3.5	3.6	3.7	3.8	3.9

Gage pressures in bottle, pounds per sq. in.

[a] Figures in this column represent the volume of carbon dioxide gas (reduced to 0° and 760 mm.) dissolved by 1 volume of water at the temperatures indicated, if the partial pressure of the carbon dioxide gas is 760 mm. Hg. Solubility data correspond to Bohr and Bock published in Landolt-Börnstein, *Physikalische-Chemische Tabellen*. Figures in the body of the table were calculated for various temperatures and pressures based on the Boyle-Mariotte law for isothermal compression.

with the aid of a device to get the gases out of the bottle or can being examined. For such methods the reader is referred to Furman [5] and to texts on gas analysis.

It may be pointed out that even though the volume in the headspace of a bottle is small, of the order of 25 milliliters, because the pressure is relatively high, as much as 5 atmospheres (equivalent to 60 pounds per square inch) in certain instances, the actual volume at standard temperature and pressure, namely, 0°C (32°F) and 760 millimeters of mercury may be as much as 125 milliliters, which is adequate for an analysis.

a. **Carbon Dioxide and Oxygen**

A modification of the usual Orsat type of gas-analysis apparatus, designed especially for the analysis of gases of carbonated beverages, is described by Toulouse.[9] It consists of two burettes and two absorption pipettes (Figure 20). One burette has a capacity of 100 milliliters and is graduated throughout its length at intervals of 0.2 milliliter. The other burette has a capacity of 15 milliliters, is 550 millimeters in length, and is graduated to twentieths of a milliliter with the graduations nearly 2 millimeters apart so that the burette can be read easily to 0.01 millimeter. The absorption pipettes are of the bubbler type. In one of these a solution of potassium hydroxide is used for the absorption of carbon dioxide and the other contains a solution of alkaline pyrogallate for the absorption of oxygen.

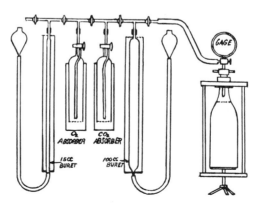

FIGURE 20. Apparatus for Examination of Gases [J. H. Toulouse, *Ind. Eng. Chem.*, 26, 768 (1934)]

Reagents—Potassium Hydroxide Solution—Dissolve 500 grams of potassium hydroxide in 1 liter of water. Do this in a hood, if available. Use care in the preparation of the reagent and in its handling. One milliliter will dissolve 40 milliliters of carbon dioxide.

Pyrogallate Absorption Solution—Weigh out 5 grams of pyrogallic acid (paper is convenient for a support) and transfer it to a reagent bottle. Dissolve 120 grams of potassium hydroxide in 100 milliliters of water and pour this solution on the pyrogallic acid in the reagent bottle. This will dissolve the acid, and the reagent is ready for use. One milliliter will dissolve 2 milliliters of oxygen.

Fill each absorption pipette to the mark with the appropriate absorption reagent.

Procedure—Make several blank determinations so that the absorbing reagents will be saturated with the gases they are not to absorb. It is necessary to absorb the carbon dioxide first in the potassium hydroxide reagent, otherwise the alkali of the alkaline pyrogallate will absorb carbon dioxide and an incorrect reading for oxygen will be obtained.

Place the bottle to be analyzed in position under the piercing mechanisms. Raise the leveling bulb so that the large burette is filled with water, making certain that the stopcocks to the pipettes are closed and the manifold inlet is open. Connect the manifold inlet to the bottle being sampled. Lower the leveling bulb to draw in the sample. Adjust the leveling bulb height so that the amount of sample can be read in the large burette. Close the inlet stopcock, open the stopcock to the potassium hydroxide pipette, and force the sample into this pipette by raising the leveling bulb. The potassium hydroxide reagent is displaced by the gas and some of it passes into the rear arm of the pipette displacing the air in that arm and pushing that air into a rubber bag. The carbon dioxide is absorbed by the potassium hydroxide solution. Lower the leveling bulb so that the gas sample is drawn again into the large burette. Repeat this process until there is no apparent change in volume. The difference between the initial volume and the final volume is the volume of carbon dioxide absorbed.

If the volume is less than 15 milliliters, arrange the stopcocks so that the gas sample can be transferred to the 15-milliliter burette. Read the volume after adjusting the water level to zero. If the volume is greater than 15 milliliters or, after reading the volume, if it is less than 15 milliliters, adjust the stopcocks so that the gas will pass into the alkaline pyrogallate solution. Raise the leveling bulb so that this is accomplished. Repeat this process until there is no further diminuation in volume. Pass the residual gas into the large burette if the volume is greater than 15 milliliters and into the 15-milliliter burette if the final volume is less than 15 milliliters. The difference between the final reading and the second reading is the volume of oxygen in the gas sample.

b. **Air in the Headspace**

A tentative method for the determination of the total air content in carbonated beverages is based on the principle of having the carbon dioxide of the beverage force the air of the headspace into a gas burette. The carbon dioxide in the gas burette is absorbed by concentrated sodium hydroxide solution, leaving the air unabsorbed. The volume of the gas remaining unabsorbed is considered to be air.

The procedure detailed in the following has been found generally acceptable for use with those carbonated beverages which contain 2.5 or more volumes of carbon dioxide. Those beverages which contain less carbon dioxide frequently do not have sufficient gas to cause the complete displacement of the air in the headspace. Such samples may be heated in a glycerol bath at a temperature of 110°C and the analysis continued until a constant air reading is obtained.

The generally accepted temperature for this type of gas analysis is 77°F. Efforts should be made to run the analysis at this temperature and to adjust the temperature of both the sample and the sodium hydroxide absorbing solution to this temperature.

The only appreciable error in the technique results from the failure to include the quantity of air that may remain in solution at atmospheric pressure. This error is eliminated if the sample is heated as described for beverages with low carbonation. It must also be taken into consideration that though the results are reported as air, complete analysis of the remaining gas might show that the amount of oxygen was less than that for normal air for if the beverage has components which may react with oxygen the oxygen content of the remaining air will be low.

Procedure—Close the needle valve of a Zahm air tester. Open the burette stopcock and place approximately 100 milliliters of water into the leveling bottle. Elevate the leveling bottle to permit the water to flow into the burette. Close the burette stopcock, lower the leveling bottle, and, after submerging the rubber seal in a small beaker containing water, open the needle valve until all the air in the piercing mechanism is displaced with water drawn from the beaker. Close the needle valve, open the burette stopcock, allow the water to return to the leveling bottle, and discard. Replace the water with a 30 per cent solution of sodium hydroxide. Raise the leveling bottle to fill the burette completely and close the burette stopcock.

Place the bottle to be tested on the base pad and push the cross bar downward to accomplish piercing of the crown. Then open the needle valve and allow gas to flow into the burette, reducing the gage pressure to about 5 pounds. Close the needle valve. Shake the apparatus a few times to allow for the absorption of the carbon dioxide in the test sample by the caustic solution and to release more gas in the beverage container. Repeat this operation until a constant air reading is obtained in the burette.

7. CAUSTIC CONCENTRATION OF BOTTLE-WASHING SOLUTIONS

Different methods are available for the determination of the caustic concentration of bottle-washing solutions. These methods may be divided into two general types. In the first of these, which is the more common, an acid or acidic substance is added to the caustic solution in the presence of a suitable indicator. In the second type, the amount of alkali is estimated conductometrically. The conductivity of an alkaline solution is proportional to the quantity of alkali present; instruments have been devised that make use of this property and values can be obtained that can be expressed in terms of percentage of caustic. Such apparatus is used together with automatic caustic feeders.

The tests in the first category can be divided into those in which tablets are used and those in which a standard acid is used to determine the amount of caustic titrimetrically.

a. Tablet Methods

The American Bottlers of Carbonated Beverages put out a kit for the determination of causticity. All of the reagents except the caustic plus aluminate test solution are in tablet form. The test is very simple to perform.

(1) Total Alkali Content

Take the sample from the soaker after it has been running for a sufficient length of time to assure thorough mixing. Allow the sample to cool to room temperature.

Procedure—Place 10 milliliters of the solution to be tested in the special A.B.C.B. 10-milliliter graduated cylinder. Add one A.B.C.B. alkali test tablet,[10] composed of potassium acid sulfate, bromophenol blue indicator, and an inert binder, crush thoroughly with a plastic or glass rod, and note whether the color of the solution is blue or yellow. If the color is yellow, the solution contains less than 1 per cent of total alkali. If blue, add and crush tablets one at a time—halves and quarters may be used to obtain greater accuracy if desired—until the mixture turns yellow. The number of tablets used is equivalent to the percentage of total alkali in the soaker solution.

(2) Caustic Content

Procedure—Place 10 milliliters of the cool sample solution into the special A.B.C.B. graduated cylinder. Add one large A.B.C.B. caustic test tablet No. 1, composed of barium chloride, tropaeolin 0 indicator, and an inert binder [11] and crush thoroughly with the plastic or glass rod. The tropaeolin indicator of the tablet determines the equilibrium point between the caustic and the tablets taken from the bottle of A.B.C.B. caustic test tablets No. 2. Note whether the color of the mixture is brick-red or yellow. If it is yellow, there is no caustic alkalinity. If the color is orange or brick-red, add 1 of the small white A.B.C.B. caustic tablets No. 2, composed of potassium acid sulfate and an inert binder [11] to the solution, crush thoroughly, mix, and again observe the color produced. Keep adding, crushing, and mixing the No. 2 tablets, one, or halves, or quarters at a time, until the mixture turns a lemon-yellow color. The number of A.B.C.B. caustic test tablets No. 2 used, divided by 2, is equivalent to the percentage of caustic alkali in the soaker solution. Each A.B.C.B. caustic tablet No. 2 is equivalent to 0.5 per cent of caustic. The percentage of caustic determined in this manner should then be checked with the caustic requirements to ascertain whether the washing solution is of the proper caustic strength.

(3) Caustic Content in the Presence of Aluminum

Soaker solutions that contain aluminates, either because they have been added to assist in bottle washing as explained in Chapter 13 or because a relatively high concentration of aluminates has been built up as a result of the solution of aluminum-foil bottle-neck labels or other aluminum-foil labels, cannot be analyzed by the unmodified A.B.C.B. caustic tablet test. The color change, in case aluminates are present, is not as sharp as one would desire; consequently, it is best to make the soaker-solution sample as clear as possible before performing the test. If suspended matter is present in the soaker solution, allow it to settle and decant the supernatant liquid or filter before running the test.

(a) A.B.C.B. Method

Procedure—Transfer 10 milliliters of the cool, properly sampled soaker solution into the special A.B.C.B. 10-milliliter graduated cylinder. Add 5 to 10 drops of caustic aluminate test solution, depending on the color intensity desired. Place the 10-milliliter graduated cylinder on a white surface, in good light, and observe the color through the side of the cylinder. The color will be yellow-green if caustic is present. Add A.B.C.B. caustic test tablets No. 2, one at a time, crush thoroughly, mix, continuing until a blue color appears and remains on stirring. Compute the caustic strength in per cent, by dividing the total number of No. 2 tablets used by 2.

Add the amount of indicator, that is, the 5 or 10 drops of caustic aluminate test solution, that experience has shown to give the color intensity most readily observed. Always place the 10-milliliter graduated cylinder on a white surface, such as a white porcelain plate or a sheet of white paper, and look at the color through the side of the cylinder. In the presence of caustic, the color will be a yellow-green, varying with the color and strength of the soaker solution. As the A.B.C.B. caustic No. 2 tablets are added, one at a time, and thoroughly crushed, the color will change to a blue and will remain blue after stirring after all of the caustic present has been neutralized. The blue color indicates that the test is complete. If there is any doubt that too many tablets have been added, the test should be repeated. In the repeated test, add the last tablet or two in halves and watch the color produced closely to determine the exact point of change.

The caustic content is calculated by dividing the number of A.B.C.B. caustic No. 2 tablets used by 2, because each No. 2 tablet is equivalent to ½ per cent caustic.

(b) Meyer Laboratory Modification

In this method [12] as in the preceding one, only the caustic No. 2 tablets are used. The tablet methods are not as accurate as the titration methods, but for some they are simpler methods to use.

Procedure—Transfer 10 milliliters of the cool soaker solution into the special cylinder or test cup. Add 5 milliliters of 20 per cent barium chloride solution, a few drops of phenolphthalein indicator solution, and stir the mixture thoroughly. Add A.B.C.B. caustic No. 2 tablets, one at a time, crushing thoroughly and mixing well after each addition with a plastic or glass rod until the solution becomes colorless. Record the number of tablets used as A.

Add ¼ teaspoonful of powdered sodium fluoride and stir. The pink color will reappear if aluminate is present, for the aluminum is sequestered as the complex sodium aluminum fluoride, $AlF_3 \cdot 3NaF$, with consequent liberation of alkali. Repeat the addition of the No. 2 tablets until the solution again turns colorless. Record the number of tablets used for the second titration as B. Subtract ½ B from A and divide the result by 2 to obtain the percentage of caustic in the soaker solution. That is,

$$\frac{A - 0.5B}{2} = \% \text{ caustic}$$

Each tablet used in the second titration, that is, B, is equivalent to 0.34 per cent of sodium aluminate or 0.45 per cent of Nalco 680.

b. **Titration Methods**

In these methods, a known volume of the sample of soaker solution is titrated with a standard acid solution to a phenolphthalein end point to obtain the caustic content and to a methyl-orange end point to obtain the total alkali. If aluminates are present, then sodium fluoride must be employed to sequester the aluminum and again a phenolphthalein end point is used to obtain the caustic content.

(1) Caustic Content

Procedure—Take a sample from the washer after it has been running for a sufficient length of time to insure that it is mixed thoroughly. Allow the sample to cool to room temperature.

Transfer 10 milliliters of the soaker solution with the aid of a pipette to a 50-milliliter titration flask. Add 5 milliliters of 20 per cent barium chloride solution and mix thoroughly. Add 2 drops of phenolphthalein indicator solution and titrate with standard 2.5 N sulfuric acid, using preferably a 10-milliliter automatic burette and adding the acid until the color of the indicator changes from pink to colorless.

With an exactly 2.5 N sulfuric acid solution, the milliliters of the titration can be read directly as percentage of caustic. Thus a titration requiring 3.2 milliliters of 2.5 N sulfuric acid is equivalent to 3.2 per cent caustic.

Great care must be taken in pipetting the caustic solution, in taking the sample, in the transfer of the sample, and in the addition of the acid. If any of the test solution or titration acid gets on the hands or clothing, it is necessary to wash the area with copious amounts of fresh water.

When the caustic solution is very dark or dirty, the test solution can be diluted and a larger flask can be used for the titration. Additional drops of phenolphthalein can also be added. If electrometric means are available for the titration, such instruments can be used. American Bottlers of Carbonated Beverages suggest that the A.B.C.B. caustic test method be used with dark solutions, for the indicator color change is more pronounced.

(2) Total Alkali

Procedure—Repeat the titration detailed before with the exception that no barium chloride solution is added and methyl orange is used as the indicator. At the end point, the indicator will change from light orange to pink.

The precautions noted in the preceding method must be observed in the determination of the total alkali also.

(3) Determination of Caustic in the Presence of Aluminate

As explained in the tablet modification, the amount of caustic present in an alkali solution containing aluminate cannot be determined by a direct titration. It is necessary to sequester the aluminum as the complex aluminum fluoride to avoid this interference.

Procedure—Transfer 10 milliliters of the properly sampled and cool soaker solution to a titration flask with the aid of a pipette. Add 5 milliliters of 20 per cent barium chloride solution and mix thoroughly. Add 2 drops of phenolphthalein indicator solution and again mix thoroughly. Titrate the test solution using 2.5 N sulfuric acid, adding the acid a few drops at a time preferably from a 10-milliliter automatic burette, until the pink color is discharged and the solution turns colorless. Record the volume of 2.5 N sulfuric acid used for this first titration as A. Do not add any more acid at this point in the determination.

Add ¼ teaspoon of powdered sodium fluoride and mix thoroughly to dissolve. If aluminates are present, the pink or rose color of the phenolphthalein indicator will again appear. Resume the titration and continue until the color of the phenolphthalein is again discharged and the solution is colorless. Record the total volume of 2.5 N sulfuric acid used in both titrations as B. The reading B includes A.

Calculate the caustic concentration by substituting the values for A and B in the following equation:

$$A - \frac{1}{2}B = \% \text{ caustic}$$

According to Cooper,[12] the number of milliliters used in the second titration multiplied by 0.68 is equivalent to the percentage of sodium aluminate present and the number of milliliters used in the second titration multiplied by 0.9 is equivalent to the amount of Nalco 680 is the soaker solution.

8. ALKALI CONCENTRATION OF COMPOUNDS

To obtain an accurate measure of the amount of caustic in a given solid mixture or in a given batch of caustic soda, it is necessary to use titrimetric or equally precise methods of analysis. A fair approximation of the alkali and caustic content of these materials can be made by use of the A.B.C.B. alkali and caustic tests.

a. A.B.C.B. Tablet Test

Procedure—Take a sample of the substance being analyzed from the center of a freshly opened container that has not been damaged or open to the air, transfer the sample to a dry jar, and close the jar. Weigh out exactly 0.5 ounce (14.2 grams) of the sample as promptly as possible so as to keep it dry. Dissolve the weighed portion in exactly 270 milliliters of distilled water (9⅛ fluid ounces).

Run the A.B.C.B. alkali and caustic determinations as detailed in section 7.a.(1) and 7.a.(2) of this chapter. The caustic and alkali concentrations of the sample can then be calculated by use of the following equations:

$$\% \text{ caustic in original material} = 20 \times \text{caustic test } \%$$
$$\% \text{ alkali in original material} = 20 \times \text{alkali test } \%$$

It is, of course, essential that the portions used for the analysis are weighed exactly.

The proportionate amount of caustic in the material can also be obtained by preparing a soaker solution, taking a sample, determining the total alkali and caustic contents, and then substituting these values in the following equation:

$$\frac{\text{caustic test, } \%}{\text{alkali test, } \%} \times 100 = \% \text{ caustic of total alkali in material}$$

b. **Titrimetric Method**

In analyzing commercial caustic soda or a mixture containing sodium hydroxide, it must be remembered that sodium hydroxide takes up water and carbon dioxide from the air. Consequently, all weighings and transfers must be done as rapidly as possible. It must also be remembered that caustic soda, if packed in blocks, first sets on the outside and most impurities are in the core of the block. It is preferable to tap different sections of the block to get a representative sample.

Procedure—Weigh a beaker and weigh into this beaker 10 grams of caustic soda or the material to be tested. Dissolve in distilled water, transfer to a 500-milliliter volumetric flask, wash the beaker thoroughly with water, add the washing to the flask, and make to volume with water. Stopper the flask, mix thoroughly, and transfer 50 milliliters with the aid of a pipette to a titration flask. Add methyl orange indicator solution and titrate with 1 N sulfuric acid to a faint-pink end point. One milliliter of 1 N sulfuric acid is equivalent to 0.031 gram of Na_2O or total alkali. In some areas, the value for 1 milliliter of N sulfuric acid is considered to be 0.032 gram of sodium hydroxide.

(1) Sodium Hydroxide Concentration

Transfer 50 milliliters of the prepared sample to a titration flask. Add 100 milliliters of 10 per cent barium chloride solution and then a few drops of phenolphthalein indicator solution, mix thoroughly, and titrate with 1 N sulfuric acid. One milliliter of N sulfuric acid is equivalent to 0.04 gram of sodium hydroxide.

(2) Sodium Carbonate Concentration

To obtain the sodium carbonate concentration, multiply the difference between the titration for total alkali and the titration for sodium hydroxide by 0.053.

(3) Direct Double Titration

As mentioned in a previous section, both free caustic and sodium carbonate can be determined on the same test aliquot by first titrating to a phenolphthalein end point and then continuing the titration to a methyl-orange end point.

Procedure—Transfer 50 milliliters of sample, containing 1 gram sodium hydroxide, to a flask, add a few drops of phenolphthalein indicator solution, and titrate with 1 N sulfuric acid as detailed in a preceding section. At the phenolphthalein end point, all of the sodium hydroxide has been neutralized and the carbonate present has been converted to sodium bicarbonate. Designate the first titration volume in milliliters of N sulfuric acid as A.

Continue the titration after the addition of a few drops of methyl-orange indicator solution. At the methyl-orange end point, all of the sodium bicarbonate has been neutralized also. Designate the volume of this titration as B.

Then the amount of sodium hydroxide can be calculated as:

$$(A - B) \times 0.04 = \text{grams of NaOH}$$

and the amount of sodium carbonate can be calculated as:

$$2B \times 0.053 = \text{grams of Na}_2\text{CO}_3$$

9. BOTTLE-RINSING EFFICIENCY TEST

In order to avoid subsequent neutralization of beverage acid, the destruction of beverage flavors and colors, or bottles of poor appearance, it is necessary to make certain that all of the alkaline bottle-washing solution is completely rinsed away from the bottles being washed.

Procedure—Add 2 or 3 drops of phenolphthalein indicator solution, prepared by dissolving 1 gram of phenolphthalein in 100 milliliters of ethyl alcohol, to a freshly rinsed bottle as it leaves the soaker. If a pink or red color is produced, all of the washing solution has not been rinsed away from the bottle. If no color is produced, the rinsing step was performed satisfactorily. In some areas, the water used for rinsing is alkaline in reaction. Such waters would normally give a positive reaction with phenolphthalein indicator solution. In such instances, an indicator that changes at a higher pH is required for the test and the indicator commonly used is thymolphthalein. Indicator solutions of this type are commercially available, for instance, from the A.B.C.B. Technical Service Laboratory.

Procedure—Add 2 or 3 drops of thymolphthalein indicator solution to a freshly rinsed bottle as it leaves the washing machine. If a blue color is produced, all of the alkaline washing solution has not been rinsed away. If no color is developed, the rinsing step was satisfactory.

10. BOTTLE UNIFORMITY

The objective of testing bottles for uniformity is to detect any marked variation in the net volume of the beverage in the bottles and to find out the proper height to which to fill the bottles to make certain that the proper net volume is maintained. The test is preferably applied to each new shipment of bottles to ascertain how the new shipment compares with the bottles in use.

a. Determination of Proper Filling Height

Procedure—Place, with the aid of a graduated cylinder, into each of 25 to 50 bottles the quantity of water equivalent to the stated capacity of the bottle or that volume of water which is equivalent to the desired volume. Measure the distance from the top of the liquid in the bottle to the top of the crown. Obtain the average of these distances and from the average calculate an index of the proper filling height to insure filling to the proper volume.

A.B.C.B. points out that the proper filling height is generally 2 to 2¼ inches below the top of the bottle for capacities of 16 ounces or less and 2¾ inches for bottles of capacities greater than 16 ounces.

b. Determination of Net Volume

Uncapping a bottle of a carbonated beverage and transferring to a graduated cylinder in order to obtain the net volume often results in a loss of liquid attributable to foaming. It is preferable, therefore, to measure the actual volume of the liquid contents by the following procedure.

Procedure—Mark the neck of the test bottle at the level of the liquid with a glass-marking pencil or crayon or paste a piece of paper or tape on the neck of the bottle and mark the level of the liquid on the paper. Uncap the bottle and pour out the contents. Fill the bottle exactly to the mark on the glass or paper with cold tap water. Transfer the water from the bottle to a 9-fluid ounce, or other appropriate volume, graduated cylinder and record the volume. This is the net volume.

11. CROWN CRIMP TEST

The American Bottlers of Carbonated Beverages gives the following details for the crown crimp test. The recommended tolerances for the outside diameter of crowned containers utilizing 26-millimeter crowns, fall between 1.130 and 1.160 inches. The allowable variation of outside locking ring diameters ranges from 1.038 to 1.060 inches. A crowner setting gage is required for the test.

Procedure—Take a crowned container from the line and apply the openings marked "No go" and "Go" for crown crimp to the crown. If both openings permit the crown to pass through, the crimp is too tight. If neither opening permits the passage of the crown, the crimp is too loose. A properly crimped crown will pass through the opening marked "Go" but will not pass through the opening marked "No go."

Remove the crown and follow the same procedure, using the openings marked "No go" and "Go" for "select bottle."

12. SPECIFIC GRAVITY

In several operations in the carbonated-beverage industry, it is necessary to determine the specific gravity of the material being handled. *Specific gravity* is the term used to express the relative masses of equal volumes of a material being measured and water at a stated temperature. The expression d_{20}^{20} referring to specific gravity is a term which represents the specific gravity of a substance at 20°C referred to water at 20°C as unity. Occasionally, specific gravity is termed relative density.

Density is mass per unit volume and is generally expressed in grams per milliliter. It is customary to refer densities to the density of water at 4°C. One milliliter of water at 4°C weighs 1 gram *in vacuo*. This is the temperature at which water has its maximum density.

The specific gravity of liquids, with which we are mainly concerned in the carbonated-beverage industry, can be measured in several ways: by hydrometer, by refractometer and reference to specific-gravity tables, by specific-gravity balance and Westphal balance, and by pycnometer. The use and application of the hydrometer has been detailed in this chapter in Sections 2.a.(1), 3.a.(2), 3.b.(2), and 3.c. The use of a hydrometer in the adjustment of the specific gravity of essential oils for the preparation of emulsion flavors was noted in Chapter 8. At times, hydrometer methods are not adequate and more accurate methods of specific-gravity determination must be employed. Two of these methods, namely, the use of the pycnometer and the Westphal balance will be described.

a. **Pycnometer Method**

A pycnometer is a small, light glass flask or tube having a definite volume. It can be used, therefore, to weigh equal volumes of liquids at a given temperature. The ratio of the weights of these equal volumes yields the specific gravity or, if the metric system is being used, the density. The most common type of pycnometer is also known as the specific-gravity bottle. There are various forms of such bottles. Some are equipped with a side arm having a capillary tube and a cap and a thermometer so that the volume of the liquid being measured can be adjusted at the desired temperature. Others consist of a small flask fitted with a ground stopper having a capillary in the stopper. In this instance, the temperature must be taken outside the flask.

Procedure—Clean, and dry the pycnometer and obtain its weight when it is empty. Fill it carefully with the liquid being measured, cooled to several degrees below the temperature at which it will be weighed, and make sure that no air is incorporated. Allow the pycnometer and its contents to come to the proper temperature either by permitting it to stand or by placing it into a bath. Wipe it dry. Remove the excess liquid from the capillary tube by means of blotting or filter paper and weigh. The temperature customarily used is 20°C.

The weight of the liquid, divided by the weight of the same volume of water, determined by calibrating the pycnometer by weighing it dry and filled with boiled, cooled water, is the specific gravity. If the specific gravity of the liquid is to be referred to water at the same temperature as that at which the measurements were made, then the following equation is used:

$$d_t^t = \frac{W_1}{W}$$

where

$d =$ the specific gravity
$t =$ the temperature designated, usually 20°C
$W_1 =$ the weight of the specific volume of liquid
$W =$ the weight of an equal volume of water

To refer this specific gravity at a given temperature to that of water at its maximum density, the right hand side of the equation must be multiplied by the factor d_W, equivalent to the density of water at the temperature t.

b. **Westphal Balance**

A simple and convenient method for the determination of the specific gravity of liquids is to use the Westphal balance or an equivalent balance. The use of this instrument is based on the principle that a body immersed in a fluid is buoyed up by a force equal to the weight of the liquid displaced. The balance comprises a stand with a leveling screw supporting a beam balanced on a knife edge. The beam has a plummet suspended from one end and counterpoised by a fixed weight with a pointer at the other end. The plummet is made of glass. It contains a thermometer and has a weighted bottom. It is constructed so that it will displace exactly 5 grams of water at 60°F. Special plummets can be made. The distance on the beam between the knife edge and the point of support for the plummet or sinker is divided into ten equal parts. The balance is equipped with five horseshoe weights or riders. Two are large and of equal weight, namely 5 grams so that one of them placed at the end from which the sinker is suspended will just counteract a buoyancy of 5 grams. The other three weights are respectively one tenth the weight of the next greater weight. These weights, when placed on the notches, that is, the points of equal division on the beam, read respectively, unit, tenths, hundredths, thousandths, and ten thousandths.

Procedure—Balance the beam in air by adjusting the setscrew. Cool the liquid to be measured, several degrees below 60°F or other designated temperature, and transfer to the test cylinder. Place the test cylinder under the point of support for the plummet and immerse the plummet in the liquid. Take care to see that no air bubbles adhere to the sinker. Place the weights in order until the beam is balanced again. Read the specific gravity at the temperature for which the plummet has been calibrated.

For example, if a plummet is being used that has been calibrated for 70°F and the specific gravity of a mixture of citrus oils and brominated oils is being determined to find out if the specific gravity of the mixture is at the point desired, as for instance 1.025 (see Chapter 8), then the first weight would be placed on the hook, the third weight on notch 2, and the fourth weight on notch 5. If the beam is balanced then the specific gravity of the oils is correct. (The second or large weight would not be used nor would the fifth weight be used unless the next decimal was being determined.) If the beam is not balanced, the correct specific gravity can be determined and the oils adjusted accordingly.

LITERATURE CITED

1. M. B. Jacobs, *Chemical Analysis of Foods and Food Products,* 3rd Ed., Princeton, Van Nostrand, 1958.
2. *Methods of Analysis, A.O.A.C.,* 8th Ed., Washington, Assoc. Offic. Agr. Chemists, 1955.
3. E. F. Langenau, in E. Guenther, *The Essential Oils,* Vol. I, New York, Van Nostrand, 1948.
4. *Standard Methods for the Examination of Water, Sewage, and Industrial Wastes,* New York, Am. Pub. Health Assoc., 1955.
5. N. H. Furman, Ed., *Scott's Standard Methods of Chemical Analysis,* 5th Ed., New York, Van Nostrand, 1939.
6. W. H. Betz and L. D. Betz, *Water Analysis,* Philadelphia, Pa.
7. *Water Analysis,* Bull. No. 11, 4th Ed., New York, Solvay Technical and Engineering Service, 1956.
8. C. A. Browne and F. W. Zerban, *Sugar Analysis,* 3rd Ed., New York, Wiley, 1941.
9. J. H. Toulouse, *Ind. Eng. Chem.,* **26,** 768 (1934).
10. U. S. Patent 1,721,809.
11. U. S. Patents 1,721,809 and 1,912,473.
12. W. C. Cooper, "Efficient Bottle Washing," Bull. No. B 139, Cudahy, Wis., Geo. J. Meyer Manufacturing Co., 1957.

SELECTED BIBLIOGRAPHY

M. B. Jacobs, *Chemical Analysis of Food and Food Products,* 3rd Ed., Princeton, Van Nostrand, 1958.

Manual of Testing Procedures, Washington, Am. Bottlers Carbonated Beverages, 1957.

W. C. Cooper, "Efficient Bottle Washing," Bull. No. B 139, Cudahy, Wis., Geo. J. Meyer Manufacturing Co., 1957.

INDEX

A

A.B.C.B. tablet tests, 311–314, 316–317
Acacia gum, 157
Acetic acid, 81–82
   acid equivalents, 87
   dissociation, 71
   ionization, 71
Acid(s), 68–88, 277
   and flavor, 111
   and yeasts, 281–282
   beverage type, 70
   concentration, 294–296
   dissociation, 71
   equivalents, 87
   ionization, 71
   price, 79
   relative strengths, 86
Acidity, 68–69, 277, 294
   in beverages, 295–296
Acidulation, 85–88
Air, 276–277, 310–311
Akwilizer, 163
Algae, 105, 286–287
Alignates, 175
Alkali concentration, 312, 315, 316–318
Alkali determinations, 311–318
Alkaline compounds, 250
Alkalinity, 93, 95, 277
   determination, 291
   total, 291–292
Alkoxyaminonitrobenzenes, 63
   toxicity, 60
Alum, 100, 102
Aluminum. See *Sodium aluminate* and *Label removal*.
Ammonium cyclohexylsulfamate, 60
Amylases, 278
Apple, 70
   composition, 240
Arabic gum, 157
Artificial flavors, 121–124
Artificial sweetners, 49–63, 302–305
   and flavor, 111

B

Bacteria, 105, 285–286
Beer carbonation, 207
Beverage(s), carbonated
   acidity, 295–296
   acidulated, 14

Beverage *(Cont.)*
 air in, 276–277
 carbonation, 207
 composition, 231–241
 definitions, 13–15
 dietetic, 49, 172
 fruit-flavored, 15, 236–241
 historical development, 16–22
 labeling, 51
 nonacid, 14
 preferences, 24–25
 production, 22–24
 sales, 22–24
 specialty, 234–236
 sugar concentration, 47–48, 301–302
Blackberry
 acid concentration, 70
 composition, 241
Body, 28, 111, 173
Bodying agents, 173–175
Boiling, 209
Bottle(s), 215–216
 dirty, 210
 filling, 221–225, 319
 machine made, 21
 net volume, 319–320
 rinsing efficiency, 318–319
 uniformity, 319–320
 washers, 251–253
 washing, 250–271
Bottlers' concentrates. See specific flavor.
Bottling, 214–230, 247
 fruit beverages, 228, 230
Brominated oils, 162–166
Brominol, 163
Brushes, 251, 253

C

Caffeine, 135
Calcite filter, 99, 103
Calcium cyclamate, 58–60
Calcium cyclohexylsulfamate, 58
Calcium hypochlorite, 96
Can(s), 216–220
 corrosion, 218
 filling, 225–228
Canning, 225–228, 229
Caramel, 180–184
Carbo-Cooler, 203–205, 245, 246
Carbohydrases, 278
Carbon adsorber(s), 99, 103–104, 105
Carbon dioxide, 21, 194–212
 and yeasts, 282–284
 determination, 200, 305–308
 liquid, 21, 198
 physiological response, 198
 pressure-temperature relationship, 208–209
 requirements, 195–197
 solid, 21, 198
 solubility, 195–197
 volume, 199–200, 305–307
Carbonated beverages. See *Beverages, carbonated* and specific flavor.
Carbonation, 200–206, 207, 305–306
 direct-service, 206–212
 loss of, 209–212
Carbonator, 201–205
 and cooler, 203–205
Cation exchange, 106
Caustic concentration, 311–316
 and aluminum, 313–314, 315–316
Caustic soda, 250–254
 concentrations, 254–259
 content, 312–316
 control chart, 258
 determination, 311–316
 predissolving, 259–264
 recovery, 267–271
 soaker solutions, 254, 255

Celery
  acid concentration, 70
  carbonation, 207
  composition, 235
  sugar concentration, 47
Cerelose, 34, 36
Chelating agents, 253
Cherry, 152–153
  acid concentration, 70
  carbonation, 207
  coloring solution, 157, 192
  composition, 240
  flavors, 152–153, 157
  sugar concentration, 47
Chlorinated lime, 96
Chlorine, 96
  available, 292–293
  excess, 93, 277
Citric acid, 71–76
  acid equivalents, 87
  dissociation, 71
  ionization, 71
  price, 79
  requirements, 72–76
  specific gravity, 73
Citrus, 25
  emulsions, 159, 161–173
Closures, 220–221
Club soda, 14, 25, 231, 232
  carbonation, 207
CMC, 174
Coagulation
  chemical, 98
  electrolytic, 104–105
Coca-Cola, 130–131, 135
  carbonation, 207
Cocoa, 13
Coffee, 13, 135
Color(s), 178–193
  certified, 185
  natural, 179–184
  solutions, 186–193
  synthetic, 184–193
  trials, 187

Coloring matter(s), 179–193
  certified, 185
  natural, 179–184
  pure-dye percentage, 185
  synthetic, 184–193
Containers, 214–220
Copper, 218, 277
Coppers, 100, 102
Coumarin, 15
Cream soda, 15
  acid concentration, 70
  carbonation, 207
  composition, 235
  flavors, 143–145
  sugar concentration, 47
Crown cap, 21, 220
Crown crimp test, 320
Crown inserts, 220–221
Crown liners, 221
Crowner, 212
Cyclamate, 56–61, 62
  ammonium, 60
  calcium, 58–60, 305
  detection, 304
  determination, 305
  formulation, 60–61, 62
  sodium, 56–57, 305
  toxicity, 60
Cylan, 56

D

Deaeration, 108–109, 277
Densitol, 163
Density, 320–321
  determination, 321–323
Dextrin, 182
Dextrose, 34
  anhydrous, 34, 36
  hydrate, 34
  sugars, 34–36
Dietetic beverages, 49, 172
Dulcin, 61, 63
  toxicity, 60

Dyes. See *Colors* and *Coloring Matters.*
pure percentage, 185

E

Electrolytic coagulation, 104–105
Emulsions, 156–173
  clear beverage, 159
  cloudy beverage, 161–173
  fruit-flavor, 157–159
  ringless, 162, 167
  specialty-flavor, 159–161
Enzymes, 278
Escherichia coli, 285
Essential oils, 116–117
Extract, 118

F

FD&C colors, 186. See also *Coloring matters, synthetic.*
Ferrous sulfate, 100, 102
Filler(s), 209, 212
Filling, 221–225, 245
  proper height, 319
Filters, 99–100, 103, 105
Flavor(s), 21, 110–124
  acid resistance, 115
  artificial, 21, 121–124
  compounding, 122–124
  criteria for, 114–116
  essence, 117
  extract, 118
  fidelity, 114–115
  fortified, 120
  fruit, 117–120, 146–154
  natural, 112, 116–120
  preferences, 24–25
  sensation, 110–112
  solubility, 114
  specialty, 21, 126–146
  true fruit, 117–120
Flavoring extract, 118

Flavoring materials, 112–114
  classification, 112–114
  natural, 112, 116–120
Flow sheets, 27, 215, 227, 229
Foaming agents, 175–176
Fruit drinks, 13, 15
  beverages, 236–241
  flavors, 117–120, 146–154
  juices, 13, 21
Fungi, 279

G

Gas analysis, 306–311
Gas volume. See *Carbonation.*
Ginger, 136–137
  extract, 139–140
  flavors. See *Ginger Ale.*
  oil, 137–138
  standards, 137
Ginger ale(s), 14, 15, 24, 25
  acid concentration, 70
  carbonation, 207
  composition, 235–236
  concentration, 62
  flavors, 135–143
  golden, 47, 70, 142, 207, 235, 236
  kola, 134–135
  pale dry, 47, 62, 70, 142, 207, 235, 236
  sugar concentration, 47
D-Glucose, 34, 182
Grape, 25, 35, 153–154
  acid concentration, 70
  carbonation, 207
  coloring solutions, 158, 192
  composition, 240, 241
  dextrose in, 35
  flavors, 153–154, 158–159
  sugar concentration, 47
Grapefruit, 171, 172
  acid concentration, 70
  bottlers' sirup, 238–239

Grapefruit (*Cont.*)
 carbonation, 207
 flavor, 172
Gravinol, 163
Gum acacia, 157
Gum arabic, 157
Gushing, 209

H

Hardness, 106
 total, 290–291
Hydro washers, 251
Hydrometer, 163, 294
 Baumé, 299
 Brix, 302
 methods, 164, 294, 298, 300–301
Hypochlorites, 96

I

Inversion, 39, 42
Invert sirup, 34
Invert sugar, 31, 32, 34, 277
Iron, 89, 93–94, 277
 total, 293
Isolates, 112–113

K

Kola(s), 14, 24, 25, 130–135
 acid concentration, 70
 caffeine in, 135
 carbonation, 207
 composition, 234–235
 concentrates, 62
 flavors, 130–135, 160
 sugar concentration, 47

L

Label removal, 264–267
Labeling, 51

Lactic acid, 80–81
 acid equivalents, 87
 dissociation, 71
 ionization, 71
 price, 79
 requirements, 80–81
Lactose, 183
Lemon, 24, 159, 169, 171
 acid concentration, 70
 and lime. See *Lime*.
 carbonation, 207
 coloring solution, 169, 192
 composition, 239
 emulsions, 159, 169, 171, 173
 flavors, 159, 169, 171, 173
 kola, 134
 sugar concentration, 47
Lemonade, 241
Lime, 24, 159, 169
 acid concentration, 70
 bottlers' sirup, 237
 carbonation, 207
 coloring solutions, 169, 171, 192
 composition, 239, 241
 emulsions, 169, 171–172
 flavors, 159, 169, 171–172, 173
 sugar concentration, 47
Lipases, 278
Liquid sugar, 31–34
 density, 299–301

M

Malic acid, 79–80
 dissociation, 71
 ionization, 71
 price, 79
Manganese, 89
Milk, 13
Mineral-water period, 16–18
Molds, 284–285

N

Normal solution, 71
Normality, 70–71

O

Oils. See *Brominated* and *Essential Oils.*
Oleoresins, 117
Orange, 24, 25, 167–168, 172
   acid concentration, 70
   beverage base, 237
   bottlers' sirup, 237
   carbonation, 207
   coloring solutions, 170, 192
   composition, 239, 240
   concentrates, 62
   emulsions, 167–168, 170, 172
   flavors, 167–168, 170, 172
   kola, 134
   sugar concentration, 47
Oxidation, 94, 276
Oxygen, 218, 308–310

P

P-4000, 63
   toxicity, 60
Palletizing, 247–248
Peach
   composition, 240
Pearson square method, 164–165
Pectin, 175, 219, 279
   demethoxylated, 175, 219
Perilla aldehyde, 64
   antialdoxime, 64
Permutit precipitator, 97–104
pH, 69–70, 277
Phosphates, 250, 254
Phosphoric acid, 82–85
   acid equivalents, 87
   dissociation, 71

Phosphoric acid (*Cont.*)
   ionization, 71
   price, 79–80
   requirements, 84–85
   specific gravity, 83
Pineapple, 168
   acid concentration, 70
   carbonation, 207
   flavor, 168
   sugar concentration, 47
Plant housekeeping, 248–249
Plant layout, 243–248
Pop, 14
Precipitator, 97–99
Premix process, 223–225, 226–228
Propoxyaminonitrobenzene, 63
   toxicity, 60
Protozoa, 105, 287

Q

Quinine water, 145–146

R

Raspberry, 124, 149–152
   acid concentration, 70
   carbonation, 207
   coloring solution, 157, 193
   composition, 240, 241
   flavors, 124, 149–152, 157
Root beer(s), 14, 15, 24, 25, 35, 159, 160, 175
   acid concentration, 70
   carbonation, 207
   composition, 235
   flavors, 126–130, 161
   sugar concentration, 47

S

Saccharin, 51–56, 302–304
"550", 50

Saccharin (Cont.)
  ammonium salt, 50–51
  detection, 302–303
  determination, 304
  formulation, 55
  requirements, 52–55
  sodium salt, 50
  soluble, 50
  sugar replacement, 55–56
  toxicity, 60
Saccharomyses cerevisiae, 280, 283
Safrole, 15, 161
Saponins, 176
Sarsaparilla, 14, 15, 126–130, 159
  acid concentration, 70
  carbonation, 207
  composition, 235
  flavors, 126–130, 161
  sugar concentration, 47
Saturator, 202, 245
Seltzer, 14, 233–234
  composition, 234
Semisynthetics, 113
Silicates, 250, 254
Sirup(s), 36–42, 43–47, 297–299
  acidified, 40–41, 42
  cold-process, 37–41
  cooler, 244
  density, 296–301
  flavored, 40
  high-density, 42
  hot-process, 41–42
  invert, 34
  room, 244
  simple, 36
  sterility, 40
Soaker(s), 251–253
  solutions, 253–259
Soda, 13, 14
  caustic. See *Caustic soda*.
  club. See *Club soda*.
  cream. See *Cream soda*.
  pop, 13, 14

Soda (Cont.)
  water, 14, 232–233
Soda ash. See *Sodium carbonate*.
Soda-water period, 18–22
Sodium aluminate, 250, 254, 263
  in washing solutions, 255, 256, 257, 313–314, 316
Sodium carbonate, 250, 254
  concentration, 318
Sodium carboxymethylcellulose, 174–175
Sodium celluloseglycolate, 174–175
Sodium cyclamate, 56–57
Sodium cyclohexylsulfamate, 56–58
Sodium hydroxide, 250. See *Caustic soda*.
  concentration, 317–318
Soft drink(s), 14
  carbonation, 207
  dietetic, 49, 172
  preferences, 24–25
  sugar-free, 172–173
Sorbitol, 62, 174
Sparkling water(s), 231–232
  carbonation, 207
  composition, 231
Specific gravity, 320–321
  adjustment, 163–165
  determination, 321–323
Spoilage, 273–287
  appearance and, 275
  bacteria, 285–286
  biochemical, 278–279
  chemical, 276–278
  chlorine, 277
  enzymic, 278
  light, 274
  microbiological, 279–287
  oxidation, 276
  physical, 273–276
  temperature, 274–275

Spoilage *(Cont.)*
  yeasts, 279–285
Stachyose, 183
Starch, 182
Sterility, 96–97, 283–284
Sterilizing agents, 96, 250
Stevioside, 64–66
Sucaryl, 56
Strawberry, 146–149
  acid concentration, 70
  carbonation, 207
  coloring solution, 157, 193
  composition, 240
  flavors, 146–149, 157
  sugar concentration, 47
Sucrose, 28–31, 32, 181, 277
  inversion, 39, 42
  standard, 29–31
Sugar(s), 26–36, 182
  and flavor, 110–111
  calories, 28
  concentration 43–47, 296–302
  dextrose, 34–36
  dilution, 43–47
  invert, 31, 32, 34
  liquid, 31–34
  standard, 29–31, 33–34
Sulfamate(s), 56, 60
Sulfur compounds, 218
Sweeteners, artificial, 49–63
Sweetening agents, 49–66
  artificial, 49–63
  natural, 63–66. See also *Sugars*.
Syncrometer, 224, 244, 246
Syn-Cro-Mix system, 223
Synthetics, 113–114

T

Tafelwasser, 231
Tartaric acid, 76–79, 277
  acid equivalents, 87
  dissociation, 71
  ionization, 71

Tartaric acid *(Cont.)*
  requirements, 76–79
  specific gravity, 77
Tea, 13, 135
Three-stage process, 221–223, 228
Tom Collins
  acid concentration, 70
  bottlers' sirup, 238
  composition, 241
  mixer base, 238
Total solids, 290
Toxicity, 15, 60
Trisodium phosphate, 250
Turbidity, 89

V

Vanilla, 143
  extract, 143–145
  flavor, 145, 159
Vichy, 231, 232

W

Washer(s), 251–253
  charging, 255–257
  Dumore, 252
Washing bottles, 250–271
  criteria for, 250–251
Washing solutions, 253–259
  strengthening, 257–259
Water, 89–109
  analysis, 289
  carbonated, 232
  classification, 90
  clear, 91–92
  color, 89
  colorless, 92–93
  criteria for, 89–90
  deaeration, 108–109
  iron in, 89, 93–94
  low-alkalinity, 95–96
  manganese in, 89

Water *(Cont.)*
  odorless, 89, 93
  organic matter in, 94–95
  purification. See *Water Treatment.*
  requirements, 90–97
  sparkling. See *Sparkling Water.*
  sterile, 96–97
  treatment, 97–109, 247

Water *(Cont.)*
  turbidity, 89
Wildness, 210

Y

Yeasts, 279–284
  acid and, 281–282
  carbon dioxide and, 282–284
  sugar and, 280–281